CONTENTS

"And changed the glory of the uncorruptible God into an image made like to corruptible man, and to birds, and to fourfooted beasts, and creeping things." ROMANS 1:23

THE MILIEU

WELCOME TO THE
TRANSHUMAN RESISTANCE

DR. THOMAS R. HORN

WITH MILIEU MEMBERS:

Dr. Frederick Meekins, Carl Teichrib, Sharon Gilbert, Troy Anderson,
Derek Gilbert, Nita Horn, Josh Peck, Wes Faull, Paul McGuire

DEFENDER

CRANE, MO

The Milieu: Welcome to the Transhuman Resistance
Defender Crane, MO 65633
©2018 by Thomas Horn with contributors. All rights reserved. Published 2018.

Printed in the United States of America.

ISBN: 9781948014038

A CIP catalog record of this book is available from the Library of Congress.

Cover illustration and design by Jeffrey Mardis.

All Scripture quotations from the King James Version unless otherwise noted

Horn's Milieu?

By Milieu Leader Dr. Thomas R. Horn

Recently, *Zygon: Journal of Religion and Science* featured a piece written by Professor S. Jonathon O'Donnell in the Department of Religion and Philosophies at the University of London titled, "Secularizing Demons: Fundamentalist Navigations in Religion and Secularity."

O'Donnell's aim? According to the article's abstract, it was to explore at a deeper level than his peers the "anti-transhumanist apocalypticisms" of our day, the central voice behind which was identified as yours truly— "evangelical conspiracist Thomas Horn [and his] milieu [community gathering place]"[1] Throughout the academic paper, O'Donnell simply refers to me and my co-obstructionists as "Horn's Milieu."

In other words, the University of London professor has determined that I and those who work with me at SkyWatch TV and Defender Publishing are the "leaders of the transhuman resistance" the members of that community had better pay attention to. The peer-reviewed *Zygon* agreed at least to the point they found reason to promulgate O'Donnell's thesis.

A title such as this placed upon me and the associates within "my milieu" is, I assure you, not as offensive as it may seem. Quite the contrary, I am encouraged that my line of work in the media has captured the attention of renowned spokespersons of the scientific world—such as those deemed worthy of being featured in *Zygon*. The very fact that personalities of such high regard and academic acumen are addressing what I've concluded throughout my career proves that my work is *worth* addressing. If I was completely off the mark out in La-la Land, I may still gain attention for being an agitator, but to persons such as O'Donnell, I would simply be considered inconsequential and not worth the precious time it takes to write such a detailed response to my convictions. Likewise, I am deeply flattered to see how professionally and fairly my stance on the subject was handled, as I nearly expected this to be yet another "hit piece" upon myself and my associates. Most naysayers who have come against our work have formulated little more than a straw man's argument. It's clear by their retaliation that they haven't truly read my work to begin with. They assume—errantly so—that we are just religious men and women who cry "sin" against anything laboratory-created, as it doesn't innately have the hand of God upon it. Such may not have been the case for O'Donnell. Much to my surprise, O'Donnell did not appear to be driven by the sole desire to paint me as a half-witted lunatic like some others have done throughout recent history. For that, I openly offer my gratitude.

There were, however, a few intrinsic flaws in his conclusions about us and our "demonologies" of transhumanism and secularism.

Religion? Science?…or Common Sense?

There is always an inherent fallacy when a personality trained in the scientifically oriented school of "show me proof/evidence" analyzes a stance based on religious conviction, particularly as it relates to eschatological prophecy and ruminations about events that have not yet occurred. It will probably always be this way on this side of eternity, since what we "evangelical conspiracists" see written on the wall regarding the scriptural warnings of God aren't *impartially* considered amidst the "proofs" and "evidences" of science. It's no secret that science recurrently discounts religion, and by extension any conclusion that a religious spokesperson may come to. Though I wouldn't immediately accuse O'Donnell of it personally, well-learned academics heavily entrenched in the scientific community or its publications quite frequently respond to religious concerns with "intellectual snobbery." And we're not guiltless on the religious end either, as many a minister will immediately regard people who place their faith in science with an air of "moral snobbery." It should not have to be this way, but it is a reality perpetuated just as much or more by the *Church* than by the secular world as I see it. Religion—especially conventional Christianity—recurrently discounts science, and though I understand why, I think this is an unnecessary tragedy. Many preachers and teachers of the Church today distance themselves from addressing scientific development and discovery (and the more "fringe" the topic, such as transhumanism, the greater the distancing), because the Bible is always the final authority, so "science is irrelevant when it contradicts Scripture"—or so the thought process relates. On the other hand, the world at large (including Christians) considers most aspects of transhumanism (and science in general) to be enormously important to our current and future generations, so it matters a great deal. How sad, then, that any minister willing to tackle the "fringe" topics is seen as a conspiracy-theorist whacko (often by the Church as well) instead of the man with the salve that could potentially help heal some of the festering tension between secular and religious people groups.

In many cases, such as a pastor of a local congregation whose duty is primarily to sharpen the Body of Christ and spread the Good News of our Savior, speaking of transhumanism and its potential travesties might be a distraction from what he or she is personally called and gifted to do. Though, there remains a massive deficit of voices on the God-fearing end of the spectrum that choose to stand in opposition to certain aspects of transhumanism and its prospective travesties, and the result of this is a tragedy in itself. When so few spokespeople of the Gospel are willing to address such a hot-button issue in the secular sphere: 1) Believers have nobody on their side to tell them *why* or *how* these scientific developments are immoral and possibly devastating, and they are therefore misled to support what they don't know the Bible expressly forbids; 2) nonbelieving agitators against religion have that much more reason to blast the world with the claim that the Church is outdated and irrelevant; and 3) those who may or may not feel convicted to believe in or follow God already (or those who have yet to even consider what they believe) have a scarce few intelligent, educated theological responders to listen to and guide them (most of whom are already squelched by a "conspiracist" label), and therefore have no reason to see Christianity as anything more than an archaic system of beliefs that can't relate to the modern world they live in.

The more a minister of the Word compares Scripture to science and the convictions and agendas therein, the more Scripture proves itself true from the beginning, so really, a minister's attention toward what is happening in laboratories across the globe today should not be as intimidating as it has become anyway—yet it appears that hardly anyone in the mainstream Church takes this position. Ministers are missing a great opportunity here to reach those who are confused about what Scripture actually says on the subject...

I don't wish to insinuate that it is every Gospel minister's responsibility to drop everything and start preaching about the inner workings of every sinister scientific organization and the impending dooms that may be birthed from them (and, as I said prior, this might prove to be a distraction away from many ministers' personal callings), but it *is* my goal in this thread of thought to point out that there have been very few ministers of the Gospel who have stood up to address these issues for all three people groups just mentioned—which covers all three known categories of people in existence (believers, nonbelievers, and those who haven't decided yet).

I can't say that I blame anyone. To take the bull by the horns and speak out regarding the concerns of modern science is a risky move, and it places one in a minority position. My opinions and concerns about the incoming "replacement humans" are often disregarded or staunchly opposed by the scientific community (and lay readers who follow it), despite that a great deal of my arguments have been solidly based on what concern-flags *the scientific community has raised in the first place.* (As for me or my associates being labeled "conspiratists," a conspiracy is not a conspiracy when there's proof, and much of my research traces back to the horse's mouth, so that label largely cancels itself out if we're all willing to be honest here…) A minister can quote from the same ten science journals in a row and identify the same impending catastrophes as a secular spokesperson on any given day, but between these two voices, by default of his religious distinction, the minister's warnings may be discarded by the mainstream who believe his conclusions to be born out of a sacred duty that many believers don't yet fully comprehend, and toward a God that nonbelievers need not heed. He or she has become a part of the minority, much like many of the prophets in the Old Testament when the majority of Israelites were so consumed with pagan revelry that the warnings of God fell to the wayside until it was too late—and seasons of great tribulation followed every time. (In the case of transhumanistic sciences, however, we are not talking about "seasons" of tribulation, but potentially a permanent and unchangeable reality that no petri dish or super-computer on the planet can reverse.)

I can certainly relate to this—*not* because I equate myself with an Old Testament prophet, but because I have attempted to raise awareness of irreversible dangers transhumanism poses to the human race that God designed, and because I have watched as, over the years, many have ignored the warnings I believe God has entrusted me to put out there. How exciting and flattering, then, that my warnings have been heeded by a man in O'Donnell's position who finds it necessary to concentrate upon and attend to the research I've compiled regarding the relationship between today's "Promethean faith" (transhumanistic ideals of "playing God" by redesigning His creation; more clarification on this later) and the clear-cut commands of God against some of these practices as addressed in the Word—even as early as Genesis.

But what of scriptural warning, anyway? Right? If a person doesn't believe in God, why would he or she care what His Word has to say about anything?

At the end of the day, it's not necessarily all about "what the Bible says" or how many Christians are informed, and I think this is a big-picture fact that most (including O'Donnell) miss when reflecting upon the work of The Milieu (hereafter capitalized to identify us as the official group O'Donnell coined). Just as often, it's a matter of *common sense* that even on a purely biological level—*removing religion and Scripture from the equation entirely*—science is changing what it means to be human, and there will never be enough genetically, artificially, or chemically engineered rats in a labyrinth to confidently show all the inconceivable ramifications of indefinitely altering the genetic makeup of *humans*.

When the rubber meets the road toward an unknown destination for humanity's future, the argument doesn't have to take place between science and religion in order for an immediate and terrifying Pandora's Box to present itself on the horizon for the human race. It doesn't take a Bible scholar to see the red flags of many modern sciences and feel concerned about the irrevocable repercussions of tampering with our human genome to the point that we no longer resemble what God *or* evolution formed us into at the onset—whichever of these two convictions one may belong to (I obviously belong to the former)— and the disastrous implications that may hold for all people on the earth. Intervention is not always driven by *religious* duty; sometimes it's driven by a preventative, humanitarian duty, and I believe most people would agree that if another life is in danger, we should all strive to be humanitarians.

Motives of The Milieu

The question then becomes: *Does* transhumanism pose a threat to the future of human life? We will spend a great deal of time unpacking that question and its potential answers in the coming pages. For now, suffice it to say that *if* transhumanism poses a threat to human life, then suddenly the label stamped upon the whistleblowers of The Milieu no longer reads "evangelical conspiracy" but "prudent forethought," and our motives are no longer related to proselytization, but to a humanitarian duty.

This is an approach that O'Donnell didn't address at all in his article. I only wish that, in all O'Donnell's complicated, grandiloquent speech (not intending to make a jab here, but he used only the most complex terminologies and word arrangements possible throughout most of his article when much of what he said could have been easily simplified), he would have shown a willingness to look a little deeper into our motives than he did. In my personal opinion, the reason he was so far off the mark in this area is because of the method he used in analyzing our common theologies. Instead of acknowledging that anyone out there including The Milieu might hold a cautionary stance against transhumanistic developments because they could be dangerous to our fragile mortality (concerns of *this current life* as it applies to everyone, regardless of religious conviction), he focuses *only* on how we interpret Scripture and what threat transhumanism holds over the immortal soul, as well as what role it plays in eschatological and societal events (concerns of the *next* life and Bible prophecy).

From the standpoint of The Milieu, it's irresponsible to separate one from the other: It's not *just* what the Bible says about the spiritual realm, and it's not *just* how all these sciences affect the future of humanity on this earth…it's *both*.

By choosing to only list what The Milieu believes about the spiritual implications of today's scientific trends, O'Donnell has whittled our message down to one that can be immediately written off by any secular mind who doesn't believe in or care about what the God of Christianity is offended by. This was the *first* glaring error in his breakdown.

As for the *second*: Many of the assessments he makes about The Milieu lack pure objectivity… For instance, he tends to cherry-pick what sticks out to him from a number of different theological discussions, threads, and voices—many of which were produced over the period of several decades—and then mesh them together into one grand-scale theological conclusion under which everyone within The Milieu answers for as a group. *Yes*, we are all united as a group, and *yes*, we have the same aware-ness-raising goals, but nobody should take a subject as complicated as religion's response to science's "transcendent man" from twenty people over twenty years, find a few common fibers, weave an all-consuming, authoritative "umbrella" doctrine from the blend…and then publish an exhibition piece blaming The Milieu for why such a garbled mess sounds irrational. We could arrive at the same irrationality if we took our favorite blips of medical advice out of twenty diet books by twenty dieti-cians over twenty years and formed a nonsensical "diet theology"; these dieticians would all be united toward seeing everyone eat healthier, but each expert would have a slightly different approach to and application for his listener or patient based on his area of study and expertise. Yet, by piecing together a grand diet plan under which the entire "medical milieu" would be accountable, we could do just as O'Donnell has done and make them appear—as a united whole—to be ignorant and fool-ish. By including bits and pieces of The Milieu's theology—thrown in a blender and funneled through a secularized worldview—O'Donnell has only shown that his argument against us lacks its own rational legs. However, because of how he articulated his position using intellectually dense (read: bombastic) words, any readers who may have approached his article with an open mind finds it tempting by the end to write off the entire Milieu community as irreconcilably conspiratorial and illogi-cal, *unless* they are perceptive enough to pick up on his marshaling tech-nique and make a conscious effort to remain objective.

For any reader of his who is willing to be intellectually shepherded by his leading, O'Donnell achieves this: 1) The Milieu cannot relate to anyone except Christians (or other religious groups), because the concerns we have for *this life* weren't brought into the discussion; 2) The Milieu cannot relate to most Christians (or other religious groups), because the theology he punctured and glued together isn't anything that anyone else can feel confident about (until they look further into our materials…but why would they, when we end up looking so crazy and conspiratorial?). Effectively, O'Donnell has whittled our message down to one that nearly nobody outside The Milieu can understand, and one that those *inside* The Milieu don't even completely agree with due to the inaccuracies of (or gaping holes *within*) his theological interpretation.

Gratitude for his reasonably professional treatment aside (he never directly came across as aggressive), the conclusions he made were often inaccurately represented, and some of them—such as the deductions he postulated regarding the motives behind our work—also rely on more than his limitedly secular and personal opinions if we are to land at a true and balanced assessment. The bias (or grand assumptions, at the very least) on his end is immediately clear.

Motive is a sticky, yet necessary, detail to address anytime someone dedicates his or her attention to the lurking beasts under glistening surfaces. Perhaps one of the most important questions one could ever ask when listening to someone labeled a "conspiritist" is: "What's motivating him [or her] to say/report these things?" If the answer to that question produces suspicion about a person's purpose or character, then that person's claims are equally suspect of being outright false, or at the least, predisposed. In the world of *science*, this is especially important, because if someone releases information while driven by an agenda, it could mislead a lot of people who believe scientific data to always represent irrefutable fact. (As any learned personality of the scientific community will tell you, scientific data is just *data*—numbers, chemical reactions, codes, and so on, often statistical—that supports other theories and hypotheses, which doesn't solidly prove anything until enough testing has been done in a specific area to produce the concrete facts. However, many conspiritists in the past [unrelated to The Milieu] have shared "data" as if it were already "facts" or "proof," but by obscuring the rest of the contributing data that possibly leads to another outcome, they're only sharing one part of the big picture, and their deductions of the situation are misleading.) Therefore, because of how important pure integrity is as it relates to agenda, the agenda and motives that The Milieu holds most important as we strive to alert the world of the dangers within the sphere of transhumanism must be herein openly admitted.

The confession you will read in this subsection is the truth, and nothing but the truth.

First of all, O'Donnell is probably correct toward the beginning of his article, when he declares that there has been a sharp increase in "conspiratorial" subcultures—and the leaders therein—who have used transhumanism as a platform to rejuvenate interest in religion. He includes myself and my Milieu contemporaries at the center of that conclusion, which is incorrect, if I might speak for myself and those associates I work with daily (and whose personal motives I am far more familiar with than O'Donnell). Several times throughout the piece, he describes these leaders (us) as "reactive" to the "marginalization" or "obsolescence" (the state of becoming obsolete) of religion,[2] which, put more simply, suggests that our goal in raising awareness about transhumanism is some crude attempt at "rescuing Christianity" (my words, not his) in an increasingly secularized era: If religion is becoming obsolete in the Western, secularized world, then religious men and women must fight to have it restored—and what better way to accomplish this than to…rail against *transhumanism*?

My logic radar just went silent. Give me a second to catch up…

My most immediate response to this is to gently inform my audience (O'Donnell included, if that is the case) that I don't think religion needs saving. As long as "God is on the throne and prayer changes things"— as my late friend Noah Hutchings of Southwest Radio Church always used to say—then "religion" (or, perhaps more accurately, the Gospel of Christ as it is spread all over the world) isn't going anywhere anytime soon, because, to quote an ancient adage of the Church: "God is in control." And even if "rescuing religion/Christianity" was what drove me, I'm intelligent enough to know that a feat such as that would not be accomplished by a man standing on an "anti-transhumanism platform."

I understand where O'Donnell is coming from with this assumption, but it *is only* an assumption. It's only logical (and obvious) that the frightening aspects of transhumanism appeal to the fragility of humanity, and the fragility of humanity is an excellent segue point from which to introduce concerns regarding what happens when human life expires (concepts of the afterlife), which is a crucial focus of many religions, including Christianity. Likewise, Christianity naturally opposes from many angles the alteration of God's original human design in the Garden of Eden, so a minister wishing to put his most promising, Gospel-spreading foot forward in a conversation with an unbelieving world might be motivated by raising awareness in an area that connects believers and nonbelievers alike as it pertains to our mutual, human fragility: Since we are all human, we at least have that one thing in common regardless of who believes in Scripture, and if there *is* a potential destruction, that presents a starting point of union for all who are concerned and involved. If we are in union toward a common, preventative goal, then we have grounds to more openly share personal convictions, which is a witnessing opportunity if there ever was one!

Using this linear breakdown as only one example, we can see a glimmer of the logic behind how an attack on secularized society might revitalize religion, if the attackers are successful...but I honestly don't think they ever would be, because that's not how it works. One single Christian sermon on the workings of God during periods of societal "religious rejuvenation" would make it clear to anyone that *true* spiritual revivals—known in history as the Great Awakenings, affecting entire towns, cities, states, and countries—*only* occur when the name of Jesus Christ is preached with passion and love...not when religion gets into a boxing ring with science. That is not to say that ministries built around responding to science aren't essential. The answers they provide are *crucial* in today's world, and The Milieu is no exception to that rule. It *is* to say, however, that no group of evangelical "conspiritists" are going to rejuvenate Christianity throughout the entire secularized West by throwing transhumanism under the bus.

First of all, one might take this theoretical assumption of O'Donnell's about The Milieu's motives as a compliment and unabashedly respond: "Why not? What ministers in our position would be anything less than thrilled to be associated to a fresh and universal form of preaching the Word?" And sure, we strive to be "fishers of men" the same as any other ministers, so if drawing the unsaved to Christ through a platform called "the transhumanism debate" was our sole motive, there really wouldn't be any shame in that.

However…

At its core, "seeing souls meet Jesus" is not the same motive as the generic "rejuvenation of archaic religion"—and I assure you, The Milieu does *not* believe that Christianity is threatened by the likes of science. When the end-game is seeing more precious souls enter the Priesthood of the Saints and inherit eternity, then by golly, call us "conspiritists for the Lord" if the outcome will ultimately be to hear, "Well done, my good and faithful servant[s]."

On that note, Paul's words in 1 Corinthians 9:19–23 state:

For though I be free from all men, yet have I made myself servant unto all, that I might gain the more. And unto the Jews I became as a Jew, that I might gain the Jews; to them that are under the law, as under the law, that I might gain them that are under the law; To them that are without law, as without law, (being not without law to God, but under the law to Christ,) that I might gain them that are without law. To the weak became I as weak, that I might gain the weak: I am made all things to all men, that I might by all means save some. And this I do for the gospel's sake, that I might be partaker thereof with you.

Depending on how one wants to interpret this verse, it might be said that, "To them that are involved in the transhumanism discussion, The Milieu became transhumanism discussers…that we might by all means save some…for the Gospel's sake." So if there are Christian ministers or counselors out there today with this precise approach, more power to you!

Again: There would be no shame in this as a sole motive for The Milieu's ministry endeavors. Nor is this the mere fight against the marginalization of religion in a secular society. O'Donnell does not present himself as any kind of expert on "soul winning for Christ," so I can clearly see he's confusing these endeavors (reaching unsaved individuals versus societal reformation), possibly because he hasn't personally come to understand Christ or His Great Commission. Yet, if a person researches religion in society at the depth that those of us in The Milieu have, it's clear that societal religion can frequently paint (and historically *has* painted) little more than a social club of smiles, potlucks, and handshakes. It doesn't produce a lasting and world-changing Great Awakening, because "societal religion" is never the same as a "personal relationship with Christ." Also, people potentially giving their heart to the Lord because they are afraid of what transhumanism threatens to destroy are doing so out of fear, and that is not a conversion that promises any longevity *or* sincerity—so why would The Milieu be after such a superficial goal? (As far as what "religious" [read: God-fearing] motives we *do* have behind our work, I will address that shortly.)

But as we mentioned earlier, there is a humanitarian outlook to all of this: Let us not assume—as O'Donnell has—that our motives aren't also to shed light on what is, at least for the duration of this earthly life, a *common-sense issue!*—once again *regardless* of religion, Christianity, or anyone's personal "higher power" belief systems. If a busload of people were parked on train tracks and the warning blare of the train's whistle hollered out from around the corner, I (and anyone in The Milieu) would run like a madman to get that bus off the tracks or assist in evacuating the people out of harm's way prior to collision. I wouldn't stand there and use the tracks as my "platform" to suddenly attempt a mass crisis conversion, preach a sermon, or rejuvenate interest in religion. In a state of catastrophic emergency, I'm going to jump into "go mode" to help as many as I can stay out of harm's way, and nothing about that is unique to "Tom Horn," "The Milieu," or any agenda other than to ensure—as anyone would—that people are spared from tragedy. If today's public mainstream is blindly parked on the train tracks of misinformation regarding transhumanistic dabbling (as I believe to be the case) while the steam engine of irreversible genetic alterations is blaring a stone's throw from shifting all of humanity into a devastating biological warfare, I don't have to be solely motivated by preaching sermons or "rescuing obsolete religions" to stop and shout for their attention.

Regardless of religion, or the "rejuvenation" thereof, warning others of danger is what any man in my well-researched position would—and should—do for the good of humanity's future.

Another repetitious point of reflection about The Milieu's motives that O'Donnell visits deeply (and in a way that suggests it to be an innate fallacy in his world of science) is our tendency to "entangle" our own bio-conservative concerns on one hand with theological concepts and our perspectives of increased Western secularism on the other.[3] As far as any attempt to reverse Western secularism: Again, God is on the throne and no man in a suit on a transhumanism platform is going to achieve this. As far as my "entangling" theology with the scientific trends that excite the public toward hidden disaster, I have no intention of refuting this, and I stand guilty as charged. It's what I am supposed to do.

But it's the "why" here that matters more than the "what" accusation. Most obviously, it's what any informed Christian is obligated to do, since "theological concepts" (i.e., the study of God's nature, His Word, His will, and by extension the "dos" and "don'ts" of our earthly existence) are at the root of everything we believe and every decision we make. If a person doesn't believe in God, then it would be understandable why he or she would consider it a natural fallacy to "entangle" concerns of science and theology together, because science is science and religion is religion: Complications arise when the two convictions clash, so many believe it would be simpler and perhaps more professional if I structured my materials to address only what is biologically relevant (for the good of human life on earth) and not what I perceive to be spiritual (eternal repercussions of current transhumanistic developments and trends).

Yet still, having been in this line of work as long as we have while our words have been broadcast to now millions of listeners, standing guilty for "entangling" science and theology should—if the responder is willing to see what's in front of him—openly link The Milieu to a motive that cannot be limited to any box called "rescuing Christianity" or "reversing Western secularization." To once again strike the issue at its core, it's more accurately about preserving the integrity of, and giving respect to, Creation as God designed it.

Whereas I completely respect the notion that religion and science are often two separate convictions, as a believer, there is no unhinging one from the other as it relates to the origin and preservation of humankind as God, Himself, created it to be. I believe God is the ultimate scientist, the First Scientist actually (as the Creator of the universe would naturally be), and therefore, I also believe that all scientific discovery is merely a reflection and a testimony of what "religion" (translation: God the Creator) initiated in the first place. It is to that authentic Creation Order I believe we owe the utmost reverence and *respect*. To dabble with the First Scientist's beautiful and artistic human DNA design to the point that His original race no longer resembles itself—with the idea that we are "improving" what was already there—is no less offensive than a would-be painter attempting to "improve" a Picasso masterpiece by smearing over the original work with a stick figure because it's "prettier."

Those who call themselves believers likely cannot—and probably *should not*—discuss man's scientific practices and the potential dangers therein without ultimately respecting the First Scientist's wishes. If I were face to face with an audience of one, I could carry on (and in the past *have* carried on) a conversation about the pros and cons of transhumanism without mentioning my Christian beliefs and "entangling" science and theology. But when it comes to my books, television and radio appearances, documentaries, etc.—in other words, my *ministry*; that arena God called me to respond to—it would be outright negligence for me to discuss the possible threats transhumanism poses to humanity without "entangling" those concepts with theology…and I can't see any justification for an apology.

At least as far as this accusation of O'Donnell is concerned, it should *never* be considered a fallacy to insert personal conviction (religious, theological, or otherwise) into a discussion about the future of human life. Scientific "facts" aside, it shouldn't be seen as a weakness, but a *strength*, to approach these matters with care and concern as they relate to each person's belief in something higher than ourselves. All scientists do the same thing each time they hover over a petri dish with thoughts of how this next experiment will positively or negatively affect their family/community, but because their personal convictions don't always inherently relate to respecting a divine Creation order or an afterlife, their concerns are justifiable…and I become the "religious conspiritist" who "entangles" science and theology.

O'Donnell's recapitulation of my own motives, as well as those in The Milieu, were not *always* inaccurate, but they *were* always subject to a bias that I can only assume to be birthed through a staunchly secular background, and by extension, they were summarized incompletely and with a misinforming slant. In a way, he ironically collected his own "data" about what drives us to act, and then dangerously skirted the edge of calling it "fact"—an error that is well associated to people who eventually earn the title "conspiritists." As a conclusion to this subsection, let me summarize more simply the motives of The Milieu: 1) to respect God first and foremost, as the First Scientist and therefore chief authority of His own design; 2) to engage the ethics in any arena where I see a threat posed to the health of mankind's future; 3) to uncover what Scripture has to say about a subject that few others in the Church are addressing, so that those whose faith might be affected by these subjects have a biblical perspective to consider; and finally, 4) to shed light on the spiritual implications of these earthly developments.

Fundamentals of The Milieu

I won't spend a great deal of time responding to O'Donnell's perceived deductions about who we are and what we believe, because the rest of the book will make much of that clear to readers. However, in his assessment, there were a few critical postulations about our relationships to each other and to fundamentalism that could raise a few eyebrows in our direction if I didn't stop for a moment to address them.

One of the first statements that O'Donnell considers fact is that the personalities making up The Milieu are "not formally affiliated," and that we congregate through the "now defunct Raiders News Network."[4] (It is for this reason, O'Donnell states, that the group was coined "The Milieu," which etymologically traces its roots within this context to the 1877 French literal reference "middle place" [i.e., meeting in the middle].) I wouldn't normally concern myself with correcting such a claim (nor will I spend much time on it here), but depending on how sensitive a reader might be to the subtle tone of condescension in O'Donnell's assessment of The Milieu's working relationship, it could be assumed that we are merely a group of amateurish squeaky wheels looking for grease over emails and broken websites and, as a conspiratorial minority, we gather in middle grounds to flatter each other. As far as the term "formal affiliation" might strictly represent the official fusion of two organizations, it is correct that there is no "legal" officiation that permanently or legally binds us all together. Yet some of the names O'Donnell listed in this group are actually my full-time employees, and as for many of the rest of them (such as the authors of this book who were listed in the Zygon paper as part of "Horn's Milieu"), our ministries have been so intertwined over the years—between my Defender Publishing House printing their original works in print, to television and radio programming—it has resulted in the joining of mass forces many times (including legal contracts) toward a common goal not the least of which are the many successful annual conferences we've carried out that have sold out year after year and reached hundreds of thousands, if not millions, of people worldwide. As far as the "now defunct Raiders News Network," in all reality, the website I have maintained originated as Raiders News Update in the early 1990s. Since then, there was only a brief stint when the site was changed to Raiders News Network, and it was within two years returned to Raiders News Update, but it has never been "defunct" at any point from its genesis; on the contrary, the organization of Raiders News will be reaching its thirtieth anniversary soon, and the network has never been more successful in its ministry reach as it is right now through SkyWatch Television.

Moving on…

One flaw in O'Donnell's logic early on is in his statement regarding The Milieu's "dual tendency" to hold that the Bible is the infallible Word of God, and the "sometimes contrary" devotion to validate that infallibility through "historical and contemporary evidences."[5] In other words—as an attempt to simplify O'Donnell's complicated and wordy summation—we: a) maintain that the God-inspired Word is incapable of error whilst we b) prove that inerrancy through materials that are *not* in the God-inspired Word. This accusation is naturally shocking, as it carries a whispered insinuation that The Milieu wouldn't find the Bible sufficiently genuine without the crutch of historical, extrabiblical, and secular evidence. It sounds even murkier when compared to the "Scripture proves Scripture" ideology centered at the core of any proper exegetical and hermeneutical practice that Christian scholars (and those in The Milieu) prioritize, which relies on the fact that the Bible is so God-inspired that it doesn't need validation outside of itself. On this, O'Donnell simply stated (as a *fact*, no less) that we are "sometimes contrary" in our stance on the Word's infallibility, and then he allowed the buck to stop there without giving any examples or added details regarding *when* we have proven to be guilty of this. I realize the size of his article did not allow for lengthy explanations at every turn—and his treatment of this accusation is so fleeting that it suggests O'Donnell considers this an issue of marginal importance anyway—but to the followers of The Milieu, it's a crucial matter: We either *do*, or we *do not*, wholly rely on the Bible in complete faith that it is inerrant and infallible in and of itself…and if we do not, a *major* stain appears on our reputation as teachers of God's Word.

Despite the lack of evidence behind his charge, however, it forces an unnecessary confusion between *personal faith* and *professional academics* to claim that The Milieu is "contrary" in our choice to study and consider both biblical and extrabiblical materials. That's quite a *leap* for a man as intelligent as O'Donnell to make…

There are two ways to approach this: 1) The Milieu is relying on extrabiblical evidence in order for our belief in scriptural inerrancy to be maintained; or 2) The Milieu is choosing not to ignore extrabiblical evidence or testimony that supports what we believe. The first assumes we can't have solid faith without "historical and contemporary evidences," while the second assumes we *utilize* the "historical and contemporary evidences" as tools to support the faith we firmly had in something already. O'Donnell really should know the difference, though he once again used steering terminology to make our approach to the "evidences" much more sinister than it really is, suggesting through his wordplay that by employing sources outside the Bible, we only "sometimes" uphold its self-evident infallibility.

When viewing this conundrum with such a narrow application as O'Donnell hinted toward, it does appear to be scandalous. However, consulting extrabiblical materials is a practice that countless apologetics ministries openly engage in and the Church has historically practiced. (Think about the leading Creation-based ministries we have today: They consider what science says, they consider what the Bible says, they consider what historical and archeological evidence proves, and they formulate a harmony of all contributing data until they arrive at what they believe to be facts regarding Creation versus evolution. None of their conclusions could have been deduced without professionally and academically "relying on historical and contemporary evidences," but all the while they openly believed that the Bible was infallible without those resources. Nothing about this is "contrary," not even "sometimes." It's simply making a choice to be thorough.)

O'Donnell is approaching this without the years of experience in ministry that I and The Milieu have, and all that implies for our listening audiences. Whether or not I and The Milieu believe Scripture to be inerrant, infallible, and God-inspired without external evidence (and I assure you, we do), we cannot assume that our listeners or readers do. As ministers, it would be remiss to continuously write "the Bible says so" in articles and books without addressing the sources *outside* the Bible that our audiences have heard about, and those sources of which might be seen as contrary to the Bible if a theological breakdown doesn't put it all in proper context. If an archeological dig were to occur a week from now that discovered ruins irrefutably identified as Solomon's temple, it would not change the fact that I and The Milieu already believed the temple was built in the precise manner the Bible outlined—but these "historical...evidences" *would* make a world of difference to our audiences.

Let it be known once and for all that I don't believe the Bible is empirically proven by extrabiblical sources. I believe the inerrant Bible is so filled with truth that I'm not surprised when sources outside of it argue for its authenticity to the degree that skeptic minds are sufficiently satisfied with the proof they need to believe as well.

While we're on the subject of The Milieu's "fundamentals," it is important to consider a perceived black-and-white-only worldview. O'Donnell is definitely *not* the first to come along and suggest that The Milieu views every force throughout history as *either* "theistic and spiritual—and thus godly—[or] atheistic and material—and thus satanic."[6] (He goes on to say that modern secularism fits the "satanic" category, according to The Milieu.) Nor is he the first to suggest that this worldview achieves a term so lofty as to be declared "The milieu's core beliefs" as he did.[7] This is yet another moment in his article where O'Donnell portrays more than just a hint of subjective partiality in his complicated, bombastic style of rhetoric. He doesn't come right out and openly proclaim that The Milieu are irrational extremists, but the all-encompassing way he boils down all our analyses and conclusions about the struggles of history (creatively reworded here as our "core beliefs") into *only* what is "godly" or "satanic" skillfully shepherds his readers into believing we're incapable of exercising balance and objectivity in our work. An underlying insinuation is that we are irrational thinkers—perhaps even impulsive—and every world issue we address (including transhumanistic science) will be coerced to fit one or the other of our cookie-cutter categories ("godly" or "satanic"), whether it's a fair assessment or not, simply because we are incapable, O'Donnell implies, of comprehending any reality between pure good and pure evil. As far as O'Donnell's rephrasing my own conclusion that transhumanism's potential "hell scenario" is being "orchestrated by a spiritual hand,"[8] there is truth to that when my words are left within the confines of their original *context*. Without the *context* that I had placed around such conclusions, which O'Donnell omitted, it might be assumed that The Milieu believes all of transhumanism is a scheme of a sinister mastermind named "Satan," and every man or woman dedicated to the field of modern science is either part of an enormous, diabolical plot to destroy us all, or they are greatly deceived and are serving the devil inadvertently.

Any follower of The Milieu's work will immediately recognize that by cramming the vast scope of our conclusions about transhumanism into two such narrow-crevice judgments (*only* good or *only* evil), O'Donnell has effectively coerced us into an unfit cookie-cutter of his own. It does not require a person to read, watch, or listen to all the media The Milieu has ever participated in to conclude: Our worldviews naturally acknowledge the diversity of humankind within the field of science (the backgrounds, motives, and convictions that ultimately drive the transhumanistic agenda, *many of which are honorable and good*) as we all take this journey toward the future together. Likewise, we consistently acknowledge that, despite our concerns, countless developments from this "Western" and "secularized" world have improved the quality of our own lives, and I will admit this even for myself. We in The Milieu would have to cripple our own rationality in order to clump all our assessments of the world and its history into the mainstream "good" or "evil."

It simply isn't that simple.

As far as the summation O'Donnell provides for The Milieu's view that Western secularism is its own ancient religion—a conspiracy of knowledge ultimately slated to recreate man in the "image of their god...Satan"[9] (the Antichrist is a far better choice of words here, though O'Donnell keeps it generic): I could spend twenty pages at this point clarifying what we believe to be ancient, intelligent forces currently stationed as the sinister support beams upon which the "Promethean faith" (or "Religion of Man," as some of my colleagues prefer) is built, but an explanation that promises to be as complicated as that is not a good fit here and now. Our convictions about that will be addressed elsewhere in this work, but suffice it to say that, as believers of the Word, we hold that the eschatological warnings within Scripture are true. Since we believe the disasters the Bible warns about *will* come to pass, we believe it's our duty to warn people when an obvious threat reveals itself in connection with those disasters. The developments that transhumanism continues toward are ripe to the point of bursting with connections to eschatological disaster, and it doesn't take a scholar to see it. If you believe the same Scripture as myself and The Milieu, then you will believe, as the Word states, that the last days hold extreme deception for most of humankind, and many will show loyalty to the Man of Sin—the Antichrist who appears on the scene working wonders and miracles.

But how will so many people fall prey to this if we've known about this coming false savior for two thousand years? What kind of lie will be *so* convincing that it leads an enormous fraction of the world's population to eternal separation from God as warned about in the book of Revelation?

Perhaps—and this is just one theory—it's the kind of deception that: a) starts slowly as a promise of a bright future in the science and medical industries, vowing that the future generations will be smarter, more efficient, healthier, and live longer (all endeavors born from spotlessly honorable motives); b) gains rapid speed after its genesis movement has been embraced, quickly becoming a nearly unstoppable force (which we're facing now); c) eventually moves faster than the population of the world can understand or keep up with so that what began as improving the human condition becomes "playing God" and nobody can see the hubris for what it is; so that d) later, when the warnings come, they will be tossed out as mere "conspiracy" as humanity celebrates the new and "improved" race, just in time for the Man of Sin to appear and introduce the mark of the Beast as an ingenious medical/scientific/technological breakthrough.

I understand how O'Donnell came to his conclusion that a central concern for The Milieu is transhumanism poising itself to eventually become—or usher in—the "image of...Satan." In many writings of our past, The Milieu have concluded that very possibility, and we still believe those scenarios to be a risk. However, taking that one rumination of ours as justification for saying our "core beliefs" *always* designate every force in history as either "theistic and spiritual—and thus godly—[or else] atheistic and material—and thus satanic" is an extremist move on his part that ironically lacks the balance and objectivity he elsewhere hopes to convey.

The road O'Donnell is traveling at this point—the destination he's attempting to arrive at in his analysis of The Milieu—is the idea that we: 1) see every historical force (i.e., global politics, shifts in world cultures, international influences, emerging technology, scientific and medical discovery) as holding its share in this ancient, satanic, humanistic religion most of The Milieu refers to as the "Religion of Man"; and therefore, 2) absolutely any step toward human transcendence above what O'Donnell coins the "original, God-given state of being," is, as he summarized, "becoming demonic" by our assessment.[10]

To state that we believe *some* avenues of transhumanism are characterized by or associated to imminent disaster and the masterminding forces of a dark and ancient enemy in sheep's clothing is not a stretch. Certainly, many of the scientific practices or goals we will address in this book—once taken too far and allowed to get out of hand (regardless of motive)—are blatantly catastrophic, and even bio-conservatives whose opinions on these matters have always been "professionally" secular in nature have been known in recent years to convert to more religious terminology ("heaven," "hell," "God," "Satan," "good versus evil," etc.) when comparing the weight of these sciences against the "unknown" of humanity's future.

However, it is grossly inaccurate to imply, as O'Donnell strongly has, that we associate any modern-day scientific or medical ingenuity to our own demonological concepts whenever a line is crossed between the human formed in a womb and the human that our current technology is able to enhance. For instance, I have no problem with a young mother choosing to enhance her baby boy's hearing by having an auditory implant installed (reasonable utilization of contemporary prosthetics)—but that is not the same thing as altering the child's humanity by fusing his DNA with a dolphin's (one animal with incredible hearing abilities, including echolocation) in an attempt to "cure" his hearing disability at the genetic level. No one would argue that prosthetics (especially neuroprosthetics) and genetic engineering are both tied to this parent category of modern technoscience called "transhumanism," yet The Milieu supports one of the options in this case and opposes the other: one *treats* his human condition, one *changes* his human condition, making him something that is no longer completely human.

And it is at this very juncture where *anyone* who believes each human has a *soul*, a unique spirit that exists beyond this earthly life—not just self-awareness, consciousness, and superior intelligence over the animal kingdom—must mix religious conviction with science. At the precise moment when a human is no longer *wholly* human, *theology cannot be divorced from the verdict that governs mankind's future.* (O'Donnell astutely refers to this as "theological bioconservativism."[11]) Because, for all the wonders that can be attributed to science and technology, nothing can explain the soul of a human being or its destination between this life and the next; *that* is a matter of faith, and faith is a concern of religion (or, as some prefer, a relationship with a higher power). *And*— for all the academic density that can be attributed to O'Donnell's intellectual thesis regarding The Milieu, he does not (and cannot) explain: 1) whether human beings have souls; 2) what happens to them after the host body is deceased; and, most important to this specific discussion, 3) what happens to the soul when the host is no longer the same human that nature (or God) produced. People in both O'Donnell's position and mine can speculate day and night, sometimes attributing our beliefs to science and sometimes to religion, but it doesn't change the fact that we don't have a soul trapped in a jar in a laboratory where we can study it and give irrefutable facts that bring both sides together in agreement.

Not every scientific, medical, or robotic development in the making right now is one that can fall into O'Donnell's "becoming demonic" synopsis as it pertains to the convictions of The Milieu. Yet, by omitting *balance* from his overview, he depicts only the extremist angle of our group and allows that extremism to bleed into our corporate character like a stamp of irrational, fundamental radicalism. After digesting "The Milieu" as O'Donnell depicted it, one might think we believe Band-Aids to be of the devil.

It's clear by O'Donnell's work as a whole that he appreciates balance and objectivity when it is science being challenged. However, it is equally clear by this work that O'Donnell doesn't allow that same balance and objectivity to be a priority when religion is taking its turn at the lectern.

What The Milieu Will Always Be

We have spent this short introductory chapter looking at what The Milieu is *not* as we have broken down a few of O'Donnell's imbalanced treatments. But as we move into chapter arguments written by actual members of The Milieu, let me state, once and for all, a dynamic aspect of what The Milieu *is*…and what it will always be.

We—believers *and* nonbelievers—need, *now more than ever*, to know the facts about transhumanism's leaps in the current age: the promising aspects of improving the quality of human life on the "pros" end, *as well as the concerning potential of negatively altering the quality of human life forever as the "con" that cannot be withdrawn.*

Anytime the waters of scientific discovery appear too clear, too refreshing, too good to be true, we need those willing to look beneath the glittering surface to uncover what predators are lurking just beyond what we can see—or just beyond what certain men in lab coats tell us is there. (In many cases, it's the proximate and learned man in the lab coat, *himself*, pressing the "red alert" button; artificial intelligence expert Hugo de Garis—celebrated researcher in the field of "evolvable hardware" and author of *The Artilect War*—comes to mind.)

We *need* "The Milieus" who approach science and technology with caution and restraint—regardless of what "conspiratist" labels those individuals or groups are handed from the opposing side. Without them and the hindsight awareness they raise, the predator is eventually guaranteed his feast, and we become the devoured.

I am as happy as a schoolboy on the playground at recess to respond to this need. My associates are ready for the challenge. The word "milieu" as a noun infers a set of surroundings, a central meeting place, or a coming together. But O'Donnell, by attaching this simple word as the corporate identification over myself and those in the same line of work, he has assisted in uniting us even further…and now we have a name. What one man might have intended to be a generic label for a minority of conspiritists flattering each other over a "now-defunct" website has become a tribute to our line of work. An accolade of honor. Individually, each of us is only one minority voice, and cry as loud as we might in the wilderness, one voice can only travel so far. Together, we form The Milieu—a collective force! And we *will always* challenge any modern developments that hold potential threat to this current life, our future, and our eternal souls.

Thank you, Mr. O'Donnell, for this marvelous opportunity to engage the world now as the Leaders of the Transhuman Resistance! Thank you for including each of us by name, listing the group in your paper as including: Thomas and Nita Horn, Rabbi Jonathan Cahn,

Chuck Missler, Michael Bennett, Gary Stearman, Sharon and Derek Gilbert, Douglas Hamp, Cris Putnam, Stephen Quayle, Michael Lake, John McTernan, Noah Hutchings, Donna Howell, Larry Spargimino, Douglas Krieger, Douglas Woodward, Paul McGuire, Fred DeRuvo, Carl Teichrib, Gary Bates, Russ Dizdar, Michael Hoggard, Terry James, Terry Cook, and Frederick Meekins, among others, many of whom will share in the chapters of this work.

NOTE: The following chapter is a draft section from Carl Teichrib's forthcoming book, *Game of Gods: The Temple of Man in the Age of Re-Enchantment*. As *Game of Gods* explores a range of intersecting issues, from philosophy and religion to world order and cultural evolution, this chapter on transhumanism contains references that correlate with other parts of his book.

Machines of Loving Grace

By Carl Teichrib

We are at the beginning of a new ideological and technological revolution in which the objectives are not physical power and control of the environment, but direct intervention into the fate of man himself.
—José Delgado.[12]

We are on the back of galloping technology, and we cannot dismount without breaking our necks.
—James Gunn.[13]

D id you see the robot?" I asked the porter as we passed in the dimly lit hallway of New York City's Empire Hotel.

He stopped and gave me a curious glance. "What are you talking about?"

"You passed right by a robot," I said matter-of-factly, motioning to a waiting group of people standing five doors down. "He's the one sitting."

The porter turned to look, peering at an Asian man seated on a wheeled, cart-like carrier.

"That's not a human," I emphasized. "It's a highly advanced robotic head on a mock body."

The gentleman stared, and stared some more. "No...that's a man." He sounded less convinced of what he believed than what he was seeing. "He must be handicapped, that's why he's sitting in the cart."

"Do yourself a favor. Walk up and look closely. You'll see it's not a person at all."

Pointing to an individual in the group who looked identical to the one sitting, I explained that the man standing was Hiroshi Ishiguro, a Japanese robotics inventor, and the "head" was his creation—a robotic double or *gemenoid* running on sophisticated software, automatically reacting to its changing environment. The porter blurted, "That can't be true!"

"Just walk by and take a look," I assured him. "It's not what you think."

Retreating down the hallway, my skeptical friend sauntered past the group, guardedly examining the person in question. Soon he was stopped, standing only a few feet away, peering closely at the life-like figure. And it was staring back at him.

Just a short jaunt from where this unusual interaction took place was an extraordinary gathering. For two days in mid-June, 2013, New York City's Lincoln Center was home to the Global Future 2045 International Congress (GF2045), a synthesis of man and machine, mind and matter, Eastern spirituality and Western secularism. Among the mix of prestigious personalities was James Martin, the tech entrepreneur and single largest benefactor to Oxford in its nine hundred-year history; Peter Diamandis, the creator of the XPRIZE and cofounder of Singularity University; Ray Kurzweil, Google's famed director of engineering; and the developer of SiriusXM Radio, Martine Rothblatt. Celebrated pioneer in artificial intelligence, Marvin Minsky, addressed the eager crowd through a video feed. Dmitry Itskov, the Russian media mogul who inspired the 2045 Initiative—the organizing entity behind the gathering—was in the spotlight both on and off the stage. Neuroscientists and consciousnesses theorists, robotic developers, posthuman thinkers, and religious leaders had assembled to explore the bounds of future human evolution.

Unlike global events I had attended before, my entrance to the GF2045 was made possible through a media pass. Magnum Veritas Productions, a documentary film company, had brought me on board as an advisor and interviewer. Our time at the Congress was therefore split between the event itself, press conferences, and conducting face-to-face interviews in the Empire Hotel.

It was an immersive and interactive experience with some of the leading minds in the transhumanist movement.

But what is *transhumanism*?

Trans-what?

An old but accurate definition can be found in the 1883 edition of *The Imperial Dictionary of the English Language*: "Transhuman (transhu'man), a. Beyond or more than human."[14] A contemporary description might sound like this: *Transhumanism is humanity's intentional evolution through science and technology.*

Lincoln Canon, then president of the Mormon Transhumanist Association, gave this definition in 2013: "Transhumanism is the ethical use of technology to expand our abilities from the human to the posthuman."[15]

Transhumanism is thus a changeover, a stepping stone, but not the final stage; it is a transition to a posthuman potential, moving beyond what we presently are. It is a future-oriented vision, one fueled by incredible scientific and technical advances and the possibilities they portend: greatly magnifying cognitive abilities, enhancing sensory input, genetic restructuring to permanently eliminate disease and weakness, finding ways to move our consciousness into a noncorruptible body, the extension of human life—to the point of immortality—and even *resurrecting the dead*.

A vast array of technologies and theoretical applications act like the carrot before the horse: Virtual and augmented reality, brain-computer interfacing and the anticipation of uploading one's mind into an artificial carrier, ubiquitous connection to the global network, cybernetics and chip implants, artificial intelligence (AI), robotics and self-replicating machines, nanotechnology, genetic manipulation, chemical switches for mood control and sharpened awareness, and cryonics for those who can afford to invest in a projected *reawakening*.[16]

Using these technologies and predicting their impact on individuals and civilization, the offering of *perfectibility*—of forging an optimal species with near infinite capacity through the works of our hands—becomes more than just a tantalizing dream. *It becomes a faith*.

The other option is to remain as we are, reside in our limitations, struggle for a few decades, and die. This is unacceptable.

Thus, science becomes salvific with hope placed in the speculations of what technology may bring. Transhumanists, those who hold this promise of techno-futures, look to the Singularity with anticipation—that hypothetical point in time when information and technology outpace humanity, forcing us to fully integrate into manageable matter. The Singularity will break our limitations of flesh and bone: Man, machine, and information will merge into a new creation. Posthumanity is the anticipated result, our evolution *beyond Man*: Übermensch.

Max More, founder of Extropy Institute, an early group discussing neo-human evolution, provided this explanation during an interview at GF2045:

> Transhumanism you can think of as, in some ways, an inheritor of the Enlightenment goals of fundamental progress in the human condition.... But it's taking it a step beyond the humanist goal of generally improving the human condition through science and technology and good will, and it's realizing that

humanity itself is limited. Because of our genetic heritage, there are certain limitations to our lifespan, to our health, to our mental well being, to our cognabilities. And it's realizing that we can take this humanist goal further and we can actually change the human condition itself.

We can use science and technology to understand the causes of aging and we can learn to eliminate those causes. It's not an unsolvable problem; it's basically an engineering problem, a scientific problem. There's nothing special about the human life span. It's just an accident; an evolutionary accident.... And why should we accept that? So really transhumanism is about taking control of our own human evolution, and deciding how long we want to live, how smart we want to be, how well modulated our emotions should be. It's really about turning our choices over to us rather than natural selection."[17]

Transhumanism is often understood as a secularist approach to unbounded progress, a humanist philosophy seeking mankind's expansion, acceleration, and the overpowering of natural limitations.

Historically it draws from an intellectual lineage stretching through modernity, and it remains a cerebral and techno-cultural movement. However, it is important to note that not all scientists and technology experts share the posthuman vision. Not too long ago, the mainstream scientific community, with reputations on the line, spurned those who believed in life extension and augmented futures.[18] But times have changed.

Transhumanism has stepped out from the fringes.

In terms of demographic and cultural identifiers, the movement is narrow and broad. It is narrow in that it is statistically and professionally predictable, as demonstrated by surveys within the transhumanist community. Variations exist from study to study; nevertheless, the following 2012 survey published by the *Journal of Personal Cyberconsciousness* presents an interesting snapshot. Regarding education and occupation,

28.8 percent and 27.6 percent hold graduate and bachelor's degrees, and computers and mathematics (27.8 percent) make up the largest fields of occupation. Most are unmarried (64.4 percent), ethnically white (85.4 percent), male (90.1 percent), and fall within a young professional spectrum (45.8 percent aged 20–29 and 21.5 percent between 30–39). The majority resides in large urban centers (43.8 percent), and less than 5 percent identify as rural.[19] Broadness is found in cultural messaging. Transhuman themes have been and are prevalent in the entertainment industry: *Avatar*, the *Star Trek* series, *2001: A Space Odyssey*, *The Matrix* and *X-Men* franchises, *Elysium*, *i,Robot*, *Gamer*, *Transcendence*, *Blade Runner*, *The Island*, *Virtuosity*, *Tron* and *Tron: Legacy*, *Surrogates*, *Lucy*, *District 9*, *Ghost in the Shell*, *Splice*, and the *Terminator* series—with technology being destroyer and savior. The online gaming world is enmeshed with transhumanist identifiers, along with frequent sacralizing patterns.[20]

Technology's impact on industries—including the medical field, aerospace, ground transportation, and agriculture—powerfully contributes to the discussion. The explosive growth of information technologies, Internet-based services, the smart connection of appliances and devices, and the data streams that seamlessly link society add to the general discourse. We find ourselves in the flow of the algorithm economy, whether or not we are aware of it, and most of us have *subjective identities*—online profiles shaped by viewing habits and the projection of *presence* through virtual social platforms.

We are living in an ever-changing, increasingly integrated matrix of information and technology.

Sociopolitical interpretations within the transhumanist movement are diverse, but stronger pulls to the political left are noteworthy. The 2012 survey revealed that 32.7 percent identified as liberal, 16.9 percent as socialist, and 4.2 percent as Marxist. Moderates made up 15.6 percent, and the libertarian approach with its limited government and free markets made up 27.4 percent, a sizeable minority when taken against the combined percentage of left-leaning transhumanists.[21] Surveys in

2013 and early 2017 by the Institute for Ethics and Emerging Technologies revealed a dominant leftist perspective.[22]

To the libertarian transhumanist, science will free us to explore individual tastes, sensibilities, and desires independent of centralized authority. In the Promised Land of techno-Marxism, the posthuman will operate in a perfect collectivism. Class division disappears under the revolution of evolutionary science; even gender and sex fade away in the hive existence.

Transhumanism has also been envisioned within a technocratic framework, the wiring of reality for optimal efficiency and connectedness: Algorithms and smart-energy calibration will order the course of neo-humanity.

No matter how civilization is rebooted, in this *shape of things to come*, "machines of loving grace" will lead us to green pastures.[23]

Philosophically, the movement is rooted in modernity's themes. Elements of Saint Simon's management-by-experts can be discerned; the positivism of Auguste Comte pokes out from under the surface. Broad overlap exists with Karl Marx.[24] The influence of Nietzsche is visible.[25] Darwin's evolutionary paradigm is magnified as holy revelation, and eugenics—the science of human betterment through genetic selection, made infamous by the Third Reich—is a contentious association that some transhumanists recognize and others spurn.[26]

"I don't see the connection between transhumanism and eugenics," Dmitry Itskov said during an interview, explaining that his transhumanism is for everybody.[27] When I queried Max More, his immediate response was, "I don't see eugenics as having much to do with transhumanism." He was willing, however, to acknowledge an inference, but emphasized, "Transhumanism is really about giving us the choice of who we want to be."[28]

Anders Sandberg, a Research Fellow at the Oxford Martin School, was willing to situate the movement within *liberal eugenics*—pursuing genetic improvement without recourse to authoritative mandates. Sandberg, who has a PhD in computational neuroscience, was part of the

European Union Enhance Project, "looking at the ethics of enhancing humans, extending lifespans, improving mood, improving intelligence, improving bodies."[29]

The dark history of eugenics causes many transhumanists to cringe at the suggestion of commonality. At the same time, notable figures outside of the community have suggested a neo-eugenics that dovetails with the movement. Robert Edwards, the British medical researcher who developed *in vitro* fertilization, frankly supported genetic intervention: "Soon it will [be] a sin for parents to have a child which carries the heavy burden of genetic disease."[30] Nobel laureate James Watson advocated a new eugenics.[31] Dr. John Glad, an expert in Russian history who devoted his retirement to eugenic considerations, took an inclusive view of human change: "Eugenicists also perceive a need to be open to genetic manipulation, machine enhancement, and even contact with beings from other planets."[32]

And why not seek human advancement through genetic alteration or by morphing with machines? Or if an extraterrestrial race offers us the keys to perfection, as has been suggested by some who claim alien contact—promising the mixture of genetic materials to produce an improved human hybrid—why not?[33] If all that exists is *material*, if reality is grounded in *naturalism*, if heaven on earth is a type of techno-modernity, then civilization's operating system is pragmatism. Why feel squeamish about methods? *The ends justify the means.*

True, the pragmatic mindset is readily detected, even celebrated, but practically and philosophically, this runs counter to some of the transhumanists I know. Most I have met believe in orienting change within ethically acceptable patterns, framing their perspectives through a moral compass. The Mormon Transhumanist Association (MTA) comes to mind, as does the fledgling Christian Transhumanist Association. I have personally interacted with founders and members of both groups, and I respect their good intentions: lifespan longevity, eradication of diseases, enhancing life quality, and striving to do good works. While I disagree with their cosmological worldview, they bring an important ethos to the conversation and invite critical and opposing voices to their platform. For this I applaud them.

However, the subject is more substantive than any single group. Advanced military applications—including enhancement programs through DARPA,[34] robotic assault vehicles,[35] and the race for national security artificial intelligence—fits within the milieu, as does your surveillance-enabled smartphone and the Web technologies we use daily. When India's Prime Minister Modi beamed himself as a life-sized, three-dimensional hologram into parts of his country during the 2014 election campaign, his *projection of presence* corresponded to transhuman models.[36] Autonomous cars and smart cities are part of the mix. Cyborgs are no longer future fantasies, as displayed in the April 2017 edition of *National Geographic*.[37] Cloning, biohacking, genome editing, and designer babies: Bioethics, traditionally focusing on abortion and euthanasia, now wrestles with complex neo-eugenic issues. Technical leaps increase at a rate that feels exponential. The public struggles not only to process the speed of innovation, but is morally drifting and contextually disconnected in an age when uncanny wisdom is needed.

As humanity turns to the digital cloud for direction, will our values mirror the algorithm of the masses? *Crowd-gnosis?* Group wisdom now dictates what is acceptable and what is not.

In his write-up for the GF2045 program, Peter Diamandis expounded on *living information*, an infinitely expanding algorithm. What he was describing is the desire for a *hive mind*:

> We humans have begun to incorporate technology inside ourselves. Humans themselves are becoming information technology. Over the last decades mankind has suddenly started changing from a loose collection of seven billion individuals to a new kind of perpetually morphing non-physical social tissue woven from densely interconnected arrays of mobile person-nodes.

In this process we—humanity—are becoming a new organism: a meta-intelligence. As a species, as this new organism, we are becoming conscious on an unprecedented new level, in a new cosmic-scale realm.

> As we are going through the metamorphosis of becoming this new meta-intelligence organism, we are going from evolution by natural selection—Darwinism—to evolution by intelligent direction. We are starting to direct the evolution of our biology and of our minds ourselves…. As we begin to liberate our thoughts, our memes, our consciousness from the biological constraints that we presently have, this will allow us to evolve far faster and ever faster.[38]

When Diamandis speaks of futures, others listen.

His work in developing space-based enterprises has elevated him as a world leader, and his XPRIZE has been a rallying point for influencers—think Larry Page, the founding CEO of Google; Ray Kurzweil; Arianna Huffington; Ratan Tata of the Tata empire of companies; and global technology strategist, Salim Ismail. In June 2017, XPRIZE allied with the United Nations International Telecommunications Union to promote artificial intelligence as a revolutionary means for global good—AI for healthcare, education, poverty alleviation, and human rights. The UN Sustainable Development Goals are now a priority for AI applications.[39]

AI is reshaping industries and services. Amazon's success is linked to its AI technologies. Artificial intelligence runs in the background of our online purchases and banking activities; self-driving cars are AI enabled; healthcare programs are utilizing AI applications. Your investment portfolio may already be managed by a robo-advisor. We hardly blink when artificial intelligence enters the mainstream.

Seeing the digital writing on the wall in the early years of the new century, Yale professor of computer science, David Gelernter, poetically envisioned a computer-saturated existence:

> The future will be dense with computers. They will hang around everywhere in lush growths, like Spanish moss. They will swarm like locusts. But a swarm is not merely a big crowd: Individuals in the swarm lose their identities; the computers that make up this global swarm will blend together into the seamless substance of the cybersphere.[40]

Could we find ourselves in a hive-mind economy where *everything* is interconnected in a global smart grid, the perfect surveillance system, buying and selling through an AI managed crypto-currency? If so, then the gnosis of the algorithm could perceivably manage humanity as a resource in the *transformational economy*. And as the vast Internet of things takes shape where everything from wearable medical devices to appliances to vehicles to *people*—every conceivable object, billions and billions of them—continually transmits data through embedded sensors, we may find ourselves in a world where AI dictates what is efficient and viable, and what is superfluous and in need of terminating. Remember, *global citizen*, you are the *product*.

Admittedly, the above paragraph is a dark vision, but as futurist Gerd Leonhard writes, "If we, today, cannot even agree on what the rules and ethics should be for an Internet of people and their computing devices, how would we agree on something that is potentially a thousand times as vast?"[41]

Transhuman-oriented technologies are visible across the gamut of society, and no single organization has a handle on the full spectrum of change. Ramifications, benefits, and unintended consequences press into the fabric of civilization. Opportunities for great good, great evil, and great confusion are before us—emboldened by the capacity of our technical creations.

Can we not just trust the experts?

Science and Society

Much to the chagrin of scientific intellectuals, technocrats, techno-elites, and transhumanists, the idea of trusting the experts is met with scorn by a sizeable percentage of the populace. Lately, the response to this disparagement has been to attach a label on those who question their narratives: *you are anti-science.*

While social media has demonstrated the power of scientifically misinformed memes, the accusation of *anti-science* is generally inaccurate. Criticism is not aimed at the *scientific method*, a standard and valuable tool of inquiry, nor is it directed at science and engineering as vital fields of human endeavor. Civilization has been tremendously blessed by the beneficial results of science and the dedication of engineers and innovators, from both the private and public sectors.

Much of the antagonism, however, rests on the ideological grandstanding of some within the scientific community, accompanied by a culture of political and economic interlock.

The dance goes something like this: Intellectual visionaries—armed with predictive models—make new claims, requiring (or demanding) public attention, along with funding and policy changes. A lobbying campaign kicks into motion. At the United Nations, action groups and academia give counsel and solidify the narrative at the global level. National governments, responding to the new pressures, invite stakeholders and experts to assist in formulating public policy. Politicians showboat on cut-and-paste proclamations and social management solutions. Budget lines are added and grants doled out; economic incentives create allegiances to the political-scientific consensus. Industries and corporations and trade associations line up with products and proposals; financial institutions monetize contracts, coordinate flow, and create markets; multilateral banking groups channel international commitments. The model has become much too big to fail. Therefore, scientific counterclaims and criticisms are downplayed while the invested narrative is continually reinforced. Globe-trotting celebrities, Leonardo DiCaprio or Al Gore or Bill Gates or Bono, pompously parade the cause; *trust the experts* is shouted from the red carpet to the podium and back to the safety of private jets and yachts. Meanwhile, new winners and losers in the marketplace are determined by edict and bureaucratic compulsion; regulations and restrictions and taxes are foisted on the public.

Trust fades to contempt.

By inserting Gore and DiCaprio, superstar eco-grifters, I have tipped my hand to the fact that I am describing the climate-change narrative. Unfortunately, the above description can be overlaid on other scientific and technical fields. Has science been ideologically weaponized for political gain, or politically weaponized for ideological revolution, or manipulated for mammon and monopoly? Aspects of science, ideology, politics, and economics are ostensibly indivisible. Will transhumanism follow suit? Garbed in a dogmatic faith-in-science and armed with loftily tantalizing promises of human perfection, the transhuman agenda cannot but find itself becoming intractably wrapped in a similar complex.

Political control of science and, conversely, the scientific validation of political agendas are nothing new. Segments of today's populace, however, are especially apprehensive of such interlock. And rightly so, for this complex has had profound repercussions in the public square and in personal lives.

Interestingly, the post-war decades of prosperity and protest—the 1950s to the early 1970s—revealed a techno-cultural tension that corresponds to our era of transhumanism. Human nature, it was argued in 1951, was on the threshold of being reconstituted due to scientific advancements and changing mental attitudes.[42] In those decades, optimism was placed in the transformative power of technology; the days of horse and buggy were still fresh in generational memory, yet transcontinental jetliners traversed the skies; satellites, spaceflights, and moon landings were in the public eye; new understandings in psychology and the discovery of double-helix DNA (deoxyribonucleic acid) rewrote professions; widespread use of telephones and televisions magically bridged distances; and the industrial employment of computers and automation reshaped manufacturing. Scientists and engineers were in high demand.[43] Rising living standards foretold greater betterment.

"Will man direct his own evolution?" asked Albert Rosenfeld, *Life* magazine's science editor in 1965. That year, *Life* published a series of articles on the theme "control of life." The September 10 cover splashed: "Audacious experiments promise decades of added life, superbabies with improved minds and bodies, and even a kind of immortality." Later in the serial, Rosenfeld wrote:

> As scientists daily edge closer to the solution of some of nature's deepest mysteries, no idea seems too wild to contemplate. Would you like education by injection? A larger, more efficient brain? A cure for old age? Parentless babies? Body size and skin color to order? Name it, and somebody is seriously proposing it…. Scientists tend to agree that some of the most exciting future developments will come out of insights and discoveries yet to be made, with implications we cannot now foresee or imagine. So we live in an era where not only anything that we can imagine seems possible, but where the possibilities range beyond what we can imagine.[44]

Reviewing the potential to tinker with DNA coding and possibly create a New Man, Rosenfeld asked: "Who is that we will appoint to play God for us?"[45] A few years later, he wrote the book, *The Second Genesis*, exploring what are essentially transhuman subjects.[46]

Uneasy about how technological advancements reposition human values, historian and educator John G. Burke penned the following in 1966: "As the tempo of change intensifies, one finds it increasingly difficult to maintain traditional modes of life and patterns of thought. Change itself, it appears, is sought as a way of life."[47]

Others, too, were expressing concern. Jacques Ellul argued that *technique* would become the guiding force in shaping the technological society, producing the "mass man."[48] C. P. Snow warned of specialization and politics, and that "trace of the obsessional" necessary in technical problem solving but problematic when applied to guiding civilization.[49] Former president of the University of Chicago, Robert M. Hutchins, critiqued the narrow scope of science and its attitudinal claim of having the corner on rational thought.[50]

Hutchins noted that specialization tends to assume its own importance and can misinterpret its place in the broader context. The forest is not missed because of the trees; it is missed because of the focus on a single branch.

Surveying the historical and modern interplay between technology and society, including themes of utopian futures and perfectibility, humanist Herbert J. Muller penned: "The real challenge remains that we do possess the technical means of doing almost anything we have a mind to, short of making angels of men."[51]

But could men be *controlled*?

A 1965 experiment suggested humanity could be manipulated. Yale professor José Delgado gave an impressive display of prowess over the mind when, standing inside a Spanish bullring, he demonstrated *cyborg control* over an aggressive animal. With a receiver wired into a bull's brain, the professor waited until the beast was in full charge and then, using his remote transmitter, sent a signal that abruptly stopped and turned the attacking animal. It was a startling display of biological management via an implant.

From the 1950s until the early 1970s, Delgado experimented with cerebral implants in cats, monkeys, rats, cattle, and humans. His work revealed that not only could physical movement be commanded by remote control, but so could moods and even feelings of euphoria. The implications were staggering. Could the mind be wired for better social integration? Do we now have the tools to shape human behavior? Are we willing to trade the illusions of personal privacy and individuality for a secure and managed existence? Have we reached the point of constructing a "psycho-civilized society"?

The Yale professor pontificated on such questions in his 1969 volume, *Physical Control of the Mind*:

> Even if we agree that individual freedom should in general bow to community welfare, we enter into a new dimension when considering the social implications of the new technology which can influence personal structure and behavioral expression by surgical, chemical, and electrical manipulation of the brain. We may tolerate the practicality of being inoculated with yellow fever when visiting Asiatic countries, but shall we accept the theoretical future possibility of being forced to take a pill or submit to an electric shock for the socially protective purpose of making us more docile, infertile, better workers, or happier? How can we decide the limits of social impositions on individual rights?[52]

"To outline a formula for the future ideal man is not easy," he wrote.[53]

The social implications of his experimental work remind me of a presentation I heard during the 9th Colloquium on the Law of the Futuristic Persons, held in late 2013 at Terasem Island.[54] To be fair, it was a *projection* of myself that attended. I interacted through an avatar, a *presence of my personality* in virtual reality (VR). Terasem Island itself is a cyberspace location, a digital land parcel with a conference theater, exhibits, meeting areas, and other constructions. As a user of virtual environments since 2009, primarily through the Second Life platform, I have frequented university lectures, church services, temples and sacred spaces, festivals at BURN2—the official Burning Man parcel—and transhumanist conferences. A transhuman event in VR makes perfect sense, as virtual reality is one of the most important technologies underscoring the posthuman ideal.

During the colloquium, one of the lecturers expounded on the ethical aspects of BCI—brain-computer interfaces—a proven technology that connects a person's brain to a computer system, prefigured in part by Delgado's work. With advancing BCI, it was explained, humanity finds itself in need of a *trans-ethics,* a new set of values to guide "networked individuals." Questions of individual freedom, security challenges, and the use of such devices for social control need to be considered. Could a wireless BCI application include directly linking the mind to a global network, and more than that, the creation of a brain-based *personal identification* mark to enable access as planetary citizens? Is this farfetched? Maybe, but current BCI innovation is moving from bulky gear to streamlined headsets to wireless capabilities, with the dream—wished by some, feared by others—of higher bandwidth, nanotechnology, and deep artificial intelligence integrating minds and machines in a cybernetic collective. Knowing that *synthetic telepathy* is already being tested, what are the ethical implications of BCI-enabled, brain-to-brain and brain-to-machine connections? "Mind reading" is becoming *science fact.*

"Do we need government intervention toward transhumanism?" our presenter asked, noting that some kind of global "skynet" will be required to manage the system: "We need global governance…this is so dangerous."[55]

We have come a long way.

Since the 1990s, critical evaluations have been added to the discussion of science and society. The dangers of what Neil Postman called being a *technophile,* those who "gaze on technology as a lover does his beloved, seeing it as without blemish and entertaining no apprehension for the future."[56] The warnings given by Douglas Groothuis in his still relevant book, *The Soul in Cyber-Space,* that we will mistake "connectivity for community, data for wisdom, and efficiency for excellence"[57]— and that the truth of God will be substituted for the ever-changing nature of our digital idols. The father of virtual reality, Jaron Lanier, has also made a powerful case for caution, arguing that individual meaning is being degraded in the visions of cybernetic totalism.[58]

Lanier comments on the cultural elevation of information:

But if you want to make the transition from the old religion, where you hope God will give you an afterlife, to the new religion, where you hope to become immortal by getting uploaded into a computer, then you have to believe that information is real and alive. So for you, it will be important to redesign human institutions like art, the economy, and the law to reinforce the perception that information is alive. You demand that the rest of us live in your narrow conception of a state religion. You need us to deify information to reinforce your faith.[59]

Important debates on technology and society are happening within the transhumanist movement. More internal criticism, however, is needed.

Tech enterprises are likewise wrestling with ethical dilemmas, and governments will soon find themselves debating difficult boundaries. A few Christian ministries and institutions have also been discussing implications, but more review is necessary. Awareness within the Christian community is generally lacking; churches need to be informed and equipped to understand the worldviews behind the movement, bringing sober realism and wisdom to the conversation. Seminaries and apologetics ministries ought to formulate biblical responses to the hope-in-technology and search for opportunities to speak into the subject. Moreover, such an approach would be internally helpful as Christians navigate the maze of concerns and changing issues.

We are on the back of galloping technology…

As innovation pushes us closer to posthuman promises, which way will the moral compass swing? When pragmatism clashes with ethical barriers, will transhuman goals be willingly tabled? How might the self-proclaimed "evolutionary imperative" configure in the posthuman worldview? Will transhumanists claim a position of Darwinian authority: that evolution demands the strongest survive, damning those incapable of enhancement? Is the vision of techno-humanity sacrosanct? If so, then Comte's Positivism and Darwinian pragmatism will be the guiding principles—science is all that matters, and evolutionary succession is the only measure of victory.

If it can be done, or *perceived* so, will it be—no matter the cost? David Gelernter thinks so: "Everything is up for grabs. Everything will change. The Orwell law of the future: Any new technology that *can* be tried *will* be."[60]

In our attempt to be a new species, will we act less than human?

For Christians and conservative individuals, other questions need be asked: Will we shun technologies that are medically beneficial or otherwise valuable because of associations with transhumanism? I hope not. Augmentation itself is not wrong; it could be argued that eyeglasses and heart pacemakers are technological enhancements. BCI can be helpful to individuals who are physically immobilized; VR platforms are useful in communication and education; computers and Internet connectivity are important tools for business and personal use. We daily use technologies linked to transhuman visions. Discernment is required to know the difference between the techno-faith that seeks to fundamentally transform mankind into an unknown quality and the helpful uses of innovation for present-day humanity. Will we use innovation and technology in ways that are good and advantageous? We have in the past and I trust we will continue doing so, even being trailblazers in scientific discovery and innovative development.

Transhumanism is far more than a zeal for science and technology, a fascination with digital tools and manageable matter; it is a social pressure cooker, a container heated by the intellectual forces of modernity.

It is also an attitude of religion.

Sacred Secularism and Mystical Materialism

The 2012 survey published by the *Journal of Personal Cyberconsciousness* shared some interesting insights regarding the transhuman community and religiosity: 81.4 percent claimed not to be religious and 9.5 percent declared themselves as Christians. Judaic and Buddhist identifiers made up 1.8 percent each, and 5.0 percent came from other religions.[61]

The preponderance of non-religion was also demonstrated when the Institute for Ethics and Emerging Technologies published their findings in 2013: 49 percent claimed to be agnostic or atheist, 17 percent viewed themselves as spiritual but not religious, Buddhists represented 11 percent, followed by a falling range of Protestant Christians, Judaism, pantheists, and pagans.[62]

Founder of the Transhumanist Party, Zoltan Istvan—who traveled across America in a coffin-shaped bus campaigning as a 2016 presidential candidate[63]—consistently reminds the transhumanist community of the dominant secularist viewpoint. During the 2014 Religion and Transhumanism Conference, Istvan asked: "Does the godless lifestyle support a transhumanist lifestyle?"[64] Affirming this position, he explained that religions bind humanity with chains of morality and therefore the atheist, unencumbered by such restraints, will be freer to think in transhuman terms. The troubling biblical message of sin and salvation, and those annoying Ten Commandments, must not hinder the progress of evolution. Traveling as presidential candidate in a coffin-bus is certainly original; however, the ideological construction of a godless New Man is anything but.

Not all transhumanists agree with Istvan's tactics or politics or his atheism, which combined, became a public relation spectacle—a lightning rod for and against transhumanism. Regardless, he is not without *faith*: faith in science, faith in technology and by extension, faith in power. Within Istvan's message is a subtle but evident flavor, a *religion without revelation.*[65]

The notion of a secular faith harkens back to Saint Simon's New Christianity and Comte's Religion of Humanity. But a more direct iteration to transhumanism comes through the eugenicist and internationalist, Julian Huxley.

Back in 1927, Huxley looked forward to a coming sacred secularity as outlined in his book, *Religion without Revelation*. This faith without recourse to a transcendent God would be predicated upon Darwinian evolution and natural power. Science would be the sacred force to control and direct said power, giving rise to an interconnected and unitary system of thought and action, resulting in growth and new life. It would be a *Sacred Reality*, a phrase he used in the struggle to express his vision.[66]

Three decades later in a collection of essays titled *New Bottles for New Wine*, Huxley described the new reality to be a unifying process of progressive evolution. Man's place in this transforming universe is that of a guiding hand, a devoted trustee, an agent of change—*an embodiment*. Our duty is to evolution.[67]

If indeed this were the case, then morality would need to be refashioned around coalescing principles and pragmatic measures. If the human population becomes too large, then to oppose control would be heinous; if cumulative knowledge is necessary for unifying processes, "then dogma is a threat, and any claim to exclusive possession of the truth or to suppression of free enquiry is immoral."[68]

Notice how Huxley lumped exclusive truth claims with the prohibition of inquiry. This is misleading. Exclusive truth does not hinder the pursuit of knowledge, but places the action in a context through which to judge the *pursuit itself* and its outcome; what is acceptable and what is discarded, what is right and wrong, what is factual and what is error.

Huxley's ideology, however, rejected absolutes:

> We must accept reality as unitary, and so must reject all dualistic ways of thinking. We must accept the fact that it is a process, and so must reject all static conceptions. The process is always relative, so we must reject all absolutes…. For this we must develop new methods of thinking.[69]

Sacred secularity therefore envisions man and evolution in a cosmic union, and our task is to embrace the non-binary, for "evolution thus insists on the oneness of man with nature."[70] From this point we can begin affirming the interdependence of the individual to the community and the cosmos. The person who accepts this paradigm "would acquire a new sense of oneness with the rest of existence."[71]

Huxley boasted: "We are perforce monists, in the sense of believers in the oneness of things, the unitary nature of reality; we see ourselves, together with our science and our beliefs, as an integral part of the cosmic process."[72]

Why is Huxley's position in *New Bottles for New Wine* important to transhumanism? First, he lays out the essential context for a scientifically justified sacred reality, a religion without revelation underscoring so much of the modern movement. Second, because it is in the pages of *New Bottles* where the term "transhumanism" comes into vogue.

Others had used the compound word before, but Huxley employed it in a way that foreshadowed magnitude. He presented a sacralizing rationale. Like a mountaineer noticing a compelling peak in the distance, being inwardly drawn to the summit before pondering its base, Huxley began his book with a pinnacle vision. The first chapter was titled "Transhumanism":

As a result of a thousand million years of evolution, the universe is becoming conscious of itself, able to understand something of its past history and its possible future. This cosmic self-awareness is being realized in one tiny fragment of the universe—in a few of us human beings.[73]

This destiny would be fulfilled by taking upon ourselves the "techniques of spiritual development," a cosmic duty to self and others resulting in the transformation of humanity:

The human species can, if it wishes, transcend itself—not just sporadically, an individual here in one way, an individual there in another way, but in its entirety, as humanity. We need a name for this new belief. Perhaps transhumanism will serve: man remaining man, but transcending himself, by realizing new possibilities of and for his human nature.

"I believe in transhumanism": once there are enough people who can truly say that, the human species will be on the threshold of a new kind of existence, as different from ours as ours is from that of Peking man. It will at last be consciously fulfilling its real destiny.[74]

Huxley's intellectual role was significant, but his was not the only voice pronouncing an ideal state of being.

The Jesuit priest and trained paleontologist, Pierre Teilhard de Chardin, had traveled similar intellectual paths to that of Julian Huxley. In fact, the two had met in 1946 and closely followed each other's work. Huxley even penned the introduction to Chardin's influential book, *The Phenomenon of Man*. In its pages, the Jesuit reminded his readers of Huxley's principle that man is "nothing else than evolution become conscious of itself."[75]

Surveying human progress, Chardin noticed the persistence of unifying structures, mechanics, and movements. A "Mega-Synthesis" was evident. The "spirit of the earth" would come as world-scale spiritual forces and the genesis of our collective mind compelled transformation. This would result in a "totalisation of the world upon itself."[76] As our individual consciousness coalesced around social principles, linked together in a global network, the earth would experience an awakening. Indeed, all consciousness would integrate into a global structure—an *earth-brain*—a state of connection and equilibrium known as the Noosphere.

Chardin noted that science, pursuing the character of evolution, will inevitably force man to look upon himself: "Man, the knowing subject, will perceive at last that man, 'the object of knowledge,' is the key to the whole science of nature." All science will eventually focus its energies on this pinnacle of evolution:

We find man at the bottom, man at the top, and, above all, man at the centre—man who lives and struggles desperately in us and around us. We shall have to come to grips with him sooner or later.[77]

And in searching for an understanding of man, science and religion will unify as we reach for our higher existence, "When we turn towards the summit, towards the *totality* and the *future*, we cannot help engage in religion."

Religion and science are the two conjugated faces or phases of one and the same act of knowledge—the only one which can embrace the past and future of evolution so as to contemplate, measure and fulfill them.

In the mutual reinforcement of these two opposed powers, in the conjunction of reason and mysticism, the human spirit is destined, by the very nature of its development, to find the uttermost degree of its penetration with the maximum of its vital force.[78]

As the very expression of evolution, humanity—wielding the energy of vital forces—will have to translate into a *new being*.

"What we see taking place in the world today is not merely the multiplication of men but the continued shaping of Man," explained Chardin in *The Future of Man*. "In one form or another something ultra-human is being born which, through the direct or indirect effect of socialisation, cannot fail to make its appearance in the near future."[79]

A religion of the future—"definable as a religion of evolution"— must therefore come to fruition: "a new mysticism, the germ of which must be recognizable somewhere in our environment, *here and now*."[80]

Chardin postulated that the ultra-human is on a convergence course with the universe itself, an Omega Point when our exalted consciousness blends with the "Cosmic Christ," the "divine in evolution." This is "an absolute direction and an absolute end."[81] The Omega Point, already percolating in our mystic hearts, *would be the full experience of man's evolutionary unity*.

How grand is the Omega Point envisioned? Mathematician Frank J. Tipler gives us a taste:

> We can say, quite obviously, that life near the Omega Point is omnipresent. As the Omega Point is approached, survival dictates that life collectively gain control of all matter and energy sources available near the Final State, with this control becoming total at the Omega Point. We can say that life becomes omnipotent at the instant the Omega Point is reached. Since by hypothesis the information stored becomes infinite at the Omega Point, it is reasonable to say that the Omega Point is omniscient; it knows whatever it is possible to know about the physical universe (and hence about Itself).[82]

Chardin peddled a mystical materialism, Huxley a sacred secularity; both are transhumanist heroes.

Although the current transhumanist community is largely agnostic or atheistic, there is interest in spirituality. Writing for *Business Insider*, Zoltan Istvan admitted that, "most transhumanists embrace some spirituality, including myself."[83]

Part of this sentiment comes from interactions with religiously oriented transhumanist groups. Another factor is the growing cultural detachment of spirituality from religion, presenting an acceptable veneer of separation from dogmas and creeds. With the postmodern slide from the hard materialism of modernity and the subsequent social desire for re-enchantment, an aesthetically grounded sense of existence is sought after. Everyone, atheists and agnostics included, are longing for a soul-felt feeling of connection or flow—spirituality without responsibility to metaphysical truth claims.

Just before the 2015 Colloquium on the Law of the Futuristic Persons, Istvan's campaign bus stopped at the Terasem "ashram" in Melbourne Beach, Florida, and the Church of Perpetual Life, a Hollywood, Florida, congregation anticipating techno-immortality. During the colloquium, he acknowledged that these interactions fostered discussions on spirituality and transhumanism, "especially the future of spirituality."[84]

Long before Istvan was on the presidential trail, however, techno-futurists and psychedelic explorers were considering the convergence of technology and spirituality. In his posthumously published book, *Design for Dying*, Timothy Leary expounded on transhuman themes and ideas of the self-made magical by science:

> Now that computer technology can personalize reality, it becomes less alien and external to use. We become one with it as a consequence of our ability to reach out and transform it.… Technology extends the boundary of self.… It is the age of the expanding person. This engenders a blurring of the material and 'spiritual' realms.… Who are you? You are boundless. Where are you? Here, there, and everywhere.[85]

Psychonaut, social visionary and author, Terence McKenna, frequently connected psychedelic mysticism with evolution via technology. "The two concepts, drugs and computers," McKenna said in a 1988 interview, "are migrating toward each other."[86]

At an Esalen Institute workshop in 1996, McKenna spoke of cyberspace and virtual reality, nanotechnology, and the bootstrapping of information to "higher and higher levels of self-reflection":

> It's our machines and our technologies that are now the major evolutionary forces acting upon us. It's not our political systems. It's these extra-sexual children, these "mind children" we have assembled out of the imagination.[87]

Technology would be a handmaid to new dimensions of *experiential information*, dissolving our differences and knitting us into a global community. Virtual reality would reveal our minds to one another, opening portals to silicon-enabled psychedelic mysticism.

Technology needs an agenda, the psychonaut believed, and this is the inner evolution of humanity with accompanying outer manifestations. To McKenna, such fundamental change would require an *archaic revival*, the resurrecting of a shamanistic paradigm for our current culture. In denouncing paganism, Christianity with its separating monotheism had stifled the flow of universal creativity, but with Christianity now on the wane, the Goddess would come alive once more. We would be transformed through our technological offerings and by an expansion of consciousness; civilization would experience a unifying synthesis through silicon and psilocybin. Mind and matter would merge in a new planetary paradigm.

Our evolutionary development would center on an ancient-future worldview: Self-enchanting and magical re-enchantment, material and spiritual, inner and outer, as above so below—*a techno-shamanistic community*. We will be priests in the temple of the cosmos, the earth an altar to the universe.

Technology as evolution and trans-spiritual subjects were hot topics at Esalen in the 1990s. The *Machine Dreams and Technoshamanism* workshops led attendees through mental romps:

Is information immortal? What unexplored realms of spirit will we encounter with amazing new forms of computer and electronic hypermedia? New myths will be needed.

New mysteries will be encountered. How can ancient models of mind such as shamanism, Hinduism, or voodoo help us comprehend silicon-based consciousness? Why are neuro-enhancers crucial tools for so many computer innovators? Can humans share an erotic relationship with a machine being? Are Earth spirits at play in fields of photons and electronic spin?

A neo-psychedelic subculture is co-evolving with new technologies to reveal glimpses of exotic futures.[88]

In 1996, technology-futurist Douglas Rushkoff offered a similar workshop at Esalen titled *Technoshamanism: Total Immersion in Spiritual Technology:*

> To wholeheartedly embrace technology may be the only way to partake in the next phase of human evolution.... Participants will experience brain machines, the Internet, computer fractals, Virtual Reality, tarot, astrology, and I Ching through computer programs, chaos math, hemispheric alignment audio, video games, and even new media.... The workshop will explore the techno-vision-quest and evaluate it as an adjunct to more traditional spiritual practices. The weekend will conclude with a techno-pagan rave dance.... Participants are encouraged to bring at least one item of technology.[89]

On the third weekend of August 1998, McKenna and VRML[90] developer, Mark Pesce—a transhumanist influenced by Chardin's writings—held their now-famous workshop at Esalen.

The duo explored the interlocking subjects of psychedelic consciousness, virtual reality and computer simulations, nanotechnology, complexity and the organization of information, and techno-mysticism. Their workshop title was representative of the oneness believed to come through the merger of man and machine: *Techno-Pagans at the End of History.*[91]

The idea of techno-paganism corresponds to the cultural transformation McKenna had been pointing to for a long time. In a 1985 interview touching on the subjects of psychedelics, cybernetics, and the "electronic shaman," he said, "We cannot travel much further with definitions of humanity inherited from the Judeo-Christian tradition."[92]

Seduced by the imagination we have bestowed upon our machinery, humanity is attempting to write a new definition of what it means to be human. We want to make ourselves in the image of our works. It is an act of sacred secularism, a journey to mystical materialism.

Perfectibility and Singularity

Themes of *perfection* underscore the human longing for order out of chaos, a reminder of *what could be.*

We long for ideal social and political relationships. We desire the perfect day, the perfect mate and perfect children, the perfect career, and perfect knowledge. Ever-higher aspirations and peak experiences may fit within this greater model, along with a host of other subjects: body images and views of biology, the perfecting of artistic endeavor, ethical and moral expectations, religious duties, ethnic and cultural traditions, and feelings of spiritual arrival. In a way this embodied desire for perfection erodes the common belief in humanity's inherent goodness. Why seek improvement if we are good already? That we *do* good things and are often well intentioned is not disputed; that we *are* good is another matter. *Imperfection is the norm.*

Transhumanism seeks to find practical paths to human perfectibility—or at least betterment that, when compared to our normal existence, appears radically superior and ultimate. As techno-pagan Mark Pesce elucidated:

> Men die; planets die, even stars die. We know all this. Because we know it, we seek something more, a transcendence of transience, translation to an incorruptible form. An escape, if you will, a stop to the Wheel.
>
> We seek, therefore, to bless ourselves with perfect knowledge and perfect will, to become as gods, take the universe in hand, and transform it in our own image, for our own delight. As it is on Earth, so it shall be in the heavens, the inevitable

result of incredible improbability, the arrow of evolution lifting us into the Transhuman, an apotheosis through reason, salvation attained by good works.[93]

Biblically, the problem of the human condition is not biological or cognitive or technical limitations, but *positional* separation from God our creator.

This *relational severance* occurred when mankind chose to pursue aggrandizement, to be more than man. It was an act opposed to the position God set on noble human existence, to be His representatives or image bearers on earth. Humanity chose, rather, to find a new identity via technique, the intentional transgressing of God's stated limitations.[94] *We would represent ourselves.* This act of disobedience immediately became inherent and is known as the problem of *sin,*[95] with moral and physical consequences plaguing every person; we naturally seek to be masters and saviors of our destiny, doing what is right in our own eyes. Whether inwardly or outwardly *I lie*, you lie; *I steal*, you steal; *I murder*, you murder; *I commit adultery and idolatry*, and you do the same. It is not that *we are one*, but that *we are all broken.*

When compared to the transcendent standards of Holy God, our natural state is immediately exposed. In the Old Testament, the prophet Isaiah tells us that even our deeds of justice—our proclaimed righteousness—are "like a menstrual cloth," and our sins sweep us away as the wind blows the withered leaf.[96] In the ancient Jewish world, such "filthy rags" were illustrative of being spiritually unclean, an appropriate picture, as menstruation cloth could be washed but would never become perfectly clean. The more such rags were hand-scrubbed, the more the fabric would degrade until it became worthless and discarded. It is an apt picture of our inability, by our own works, to attain *positional perfection*—to be viewed by God as righteous.

In the New Testament, the Apostle Paul reminds us that Jews and Greeks, all mankind, has fallen under the curse of sin. In reiterating the words of the psalmist, Paul points to the true state of our hearts:

The LORD looks down from heaven upon the children of men, to see if there are any who understand, who seek God. They have all turned aside, they have together become corrupt; There is none who does good, no, not one.[97]

Positional perfectibility remains out of our hands; because of our sinful nature, we are incapable of fixing this dire situation on our own. In fact, that is the point. Yet we long for perfectibility, a return to our previously unfallen state but without recourse to God's exclusive mandate: *salvation through Jesus Christ alone.*[98] So we continually seek ways to save ourselves and reclaim Eden for our own greatness.

Looking back over the history of technology and religious thought, activist Professor David F. Noble made an astute connection:

Over time, technology came to be identified more closely with both lost perfection and the possibility of renewed perfection, and the advance of the arts took on new significance, not only as evidence of grace, but as a means of preparation for, and a sure sign of, imminent salvation.[99]

Perfectibility is an inescapable feature of religious philosophy.

Confucianism seeks *self-attained perfection* through the deliberate engagement of ethical character development, right duty and actions, virtue through education, and social etiquette. Or consider Hinduism as a process, a quest for perfection in the undifferentiated Atman; or the steps of perfection, transformation, and the going beyond as a Great Being in the Buddhist tradition; or the Sufi's journey to union through the passing of stages, becoming Perfect Man in the flow of God. The Mormon faith emphasizes perfection, impelling the believer to strive for exaltation through proper behavior, temple requirements, and priesthood duties.

The examples in the above paragraph are paths of *practical perfection*: Do these things, focus on these principles, act this way, or follow these spiritual techniques. Salvation is dependent on your active participation.

Esoteric beliefs add another dimension to the perfection theme. The Hermetic Order of the Golden Dawn, arguably one of the most influential occult groups in the twentieth century, encouraged its members through rituals and experimentation "to be more than human, to transcend physical limitations"—"to be more than human, and thus gradually raise and unite myself to my Higher and Divine Genius."[100] A spiritually cryptic version of evolution promised a path to ascension and perfection, degree by degree.

Presenting a Masonic interpretation, W. L. Wilmshurst penned the following in his classic, *The Meaning of Masonry*:

> From grade to grade the candidate is being led from an old to an entirely new quality of life. He begins his Masonic career as the natural man; he ends it by becoming through its discipline, a re-generated perfected man. To attain this transmutation, this metamorphosis of himself, he is taught first to purify and subdue his sensual nature; then to purify and develop his mental nature; and finally, by utter surrender of his old life and losing his soul to save it, he rises from the dead a Master, a just man made perfect....
>
> This—the evolution of man into superman—was always the purpose of the ancient Mysteries, and the real purpose of modern Masonry is, not the social and charitable purpose to which so much attention is paid, but the expediting of the spiritual evolution of those who aspire to perfect their own nature and transform it into a more god-like quality. And this is a definite science, a royal art.[101]

Henry C. Clausen, while Sovereign Grand Commander of the Supreme Council of the Scottish Rite of Freemasonry, hinted at a coming techno-spirituality:

Science and philosophy, especially when linked through mysticism, have yet to conquer ignorance and superstition. Victory, however, appears on the horizon. Laboratory and library, science and philosophy... outstanding technicians and theologians are now uniting as advocates of man's unique quality, his immortal soul and ever expanding soul.[102]

Theosophy with its blend of Eastern religions and Western occultism, taught that perfectibility could be achieved through directed, spiritual evolution. Those who have so advanced are known as "Adepts or Supermen."[103]

According to C. W. Leadbeater, an early theosophical authority, these Supermen maintained physical bodies far exceeding normal men in terms of longevity and capacity. It was also claimed that these Adepts could transition from body to body, temporarily possessing others to achieve a purpose.

Leadbeater presented a general argument for these Perfected Men:

The existence of Perfected Men is one of the most important of the many new facts which Theosophy puts before us. It follows logically from the other great Theosophical teachings of karma and evolution by reincarnation. As we look round us we see men obviously at all stages of their evolution—many far below ourselves in development, and others who in one way or another are distinctly in advance of us. Since that is so, there may well be others who are very much further advanced; indeed, if men are steadily growing better and better through a long series of successive lives, tending towards a definite goal, there should certainly be some who have already reached that goal.[104]

This evolution into Perfect Men, it was believed, is part of the drama of incorporation into the Universal Over-Soul, the "great unit consciousness" or *Brahma* in which the oneness of all exists.[105]

The transformation of religion and society through a worldizing process is considered another dimension of this same Great Work. Universal consciousness and human evolution, rising from the lower to the higher, provides the background for an obscure statement in an 1891 edition of *Lucifer*, a theosophical magazine. In this text, the transmutation was described as passing from "the lower kingdoms of nature, up to the divine trans-human realisation at the close."[106] Roughly twenty years later Russian theosophical thinker, P. D. Ouspensky, described "cosmic consciousness" as "trans-humanizing a man into a god."[107]

What was being described from an esoteric position was a type of Singularity, a point when spiritual technique ushers in a planetary transformation, and a time when man ceases being human on the evolutionary path. Such a pinnacle moment has also been portrayed as a great *convergence* or *emergence*. The transhuman version, however, sees this primarily as a technical and informational convergence: Machine intelligence exceeding human capacity, which will spur an exponential rate of artificial cognition and force humanity to transform into a Great Being. But debate within the transhuman camp continues as to what the Singularity will be like and what it entails. It is a future oriented prediction with passionate believers and questioning skeptics. Nevertheless, it fits with Chardin's idea of the Omega Point, that supposed period when science and spirituality blossom into the evolution of cosmic consciousness and the arrival of the ultra-human.

Google's Ray Kurzweil, a rock star in the transhumanist community, correlates the Singularity to a human-machine transcendence. Kurzweil is religiously agnostic and although his approach is prophetically scientific, his concept amounts to a spiritual translation through technology. The posthuman journey approaches a *god-point*, producing an existence that would appear deified when compared to our present situation. A few lines from his *New York Times* bestselling book, *The Singularity Is Near*, helps us connect the dots behind his big idea:

As a consummation of the evolution in our midst, the Singularity will deepen all of these manifestations of transcendence....

The matter and energy in our vicinity will become infused with the intelligence, knowledge, creativity, beauty, and emotional intelligence (the ability to love, for example) of our human-machine civilization. Our civilization will then expand outward, turning all the dumb matter and energy we encounter into sublimely intelligence—transcendent—matter and energy. So in a sense, we can say that the Singularity will ultimately infuse the universe with spirit.[108]

Zoltan Istvan passionately expressed his thoughts on the Singularity to those attending the 10th Colloquium on Terasem Island:

I am a person striving to achieve and reach the Singularity.... Who doesn't want to know what the Singularity really is? Who...wouldn't press a big red button and say "I want to go there right now" and discover everything there is that far into the future; whether it's omnipotence, whether it's just omniscience, whether it's a perfect hive mind...whether it's a complete immersion into one, single entity where, one doesn't even recognize himself anymore.[109]

I believe the *unexplainable idea* of the Singularity and the transhuman quest to reach this fuzzy future is flawed.

Can it be attained or even recognized if we are unsure of what it is, especially something apparently so vast and potent and all-encompassing? Reality cannot be condensed into an algorithm, let alone a formula that equals or surpasses the human experience; machines—even artificial intelligence—remains locked in utilitarian functions, and to push beyond intended capabilities often results in damage and degradation; information is not the same as life; we are unable to define what *thought* is, and understanding consciousness remains out of reach.

Consciousness and so much of the human experience cannot be qualified or quantified in the material. To be human is more than just occupying "meat space." There is an immaterial side to existence: our spirits and souls, and the free will associated with relationships, beliefs, values, and our unique personalities. We are not material entities only, but a complex of biology, spirit and soul. *We are fearfully and wonderfully made.*[110]

Dressed in material mysticism—that it is our destiny to somehow awaken the material universe with our material proficiency—the Omega Point or Singularity is an attempt to sell an accelerating technological future based on a model from a non-existent past. It aims to translate what has never been proven: the evolutionary change from one species into an entirely new species, and not just any Darwinian transformation but the arrival of self-deified man through the perfecting of information.

But might we encounter a type of Singularity, the unification of technologies, religions, politics and ideologies, all squeezed into a world system? This would only be possible under the right international conditions of extreme conflict or social tension, providing a planetary *gestalt* experience and the opportunity to fundamentally restart civilization. The person or entity that pulled off such a feat would be hailed as a world savior, a guiding force to administer evolution. The Singularity would be embodied as the image of the New Man.

Historical patterns give some latitude to speculate as to what an ultimate *image of man* may look like. Models could include an enhanced United Nations or a replacement institution with far-reaching powers, or a "planetary king" as suggested by Yermentay Sultanmurat.[111] Theosophists anticipate a coming perfected man, the "world teacher" who is to coalesce all things in the "great work." Maybe a self-proclaimed higher intelligence from another dimension would be the *ambassador of oneness*, an extraterrestrial messiah laying claim to the evolutionary imperative by announcing that *his race* seeded mankind on earth. Or possibly the image is something we manufacture to be subsequently high jacked; an artificial intelligence possessed and "made alive" by a malevolent spiritual entity. All of this is conjecture, the stuff of movies. The Bible, however, describes a coming Man of Lawlessness.[112]

Using the vernacular of psychological warfare, could it be that the Singularity is a spiritual PSYOPS, the last great deception reflecting the first great deception?

The future, grounded in the distant past, races into our present.

Techno-Religiosity

Modern transhumanism has resurrected a marginalized philosophical religion, birthed a "trans-religion," and energized a subset of the Mormon faith.

For transhumanists in Russia, the ideas of Nikolai Fedorov and cosmism have been renewed. Fedorov, who died in 1903, was a quiet librarian who advocated a synthesized philosophy constructed from Russian Orthodoxy and technical progress. Looking upon the resurrection of Jesus Christ as an example to follow, Fedorov held that God intended humanity to engage in *self-resurrection*. In other words, if Jesus could rise from the dead, then mankind must find a technical way to likewise achieve this eminent goal. Given enough time, everyone from the past, present, and future would be raised to immortality through the spiritual application of technology.

David F. Noble merges his commentary with Fedorov's thoughts:

Federov combined the ideals of Russian Orthodoxy, the Russian aristocracy, and the Russian peasant commune into a doctrine of what he called "the Common Task," the unification of all humanity and the "removal of all the obstacles that prevented the evolution of man's humanity toward its last stage, the stage of self-creation, immortality, and God-likeness."... This transformation, which entailed the reconstitution of the bodies of past humans, demanded mankind's complete mastery and control over the universe, including space.[113]

Early Russian cosmists also incorporated theosophical and occult ideas, perennial philosophy and gnostic approaches, self-organizing complexity and chaos theory, Russian cultural sentiments, and mystical science—"higher magic partnered to higher mathematics."[114]

While Fedorov discouraged some of the esoteric teachings of his contemporary futurists, the cosmist worldview opened wide the doors of philosophical imagination. Cosmism also fit within the Russian tendency for totalitarian solutions. In his book, *The Russian Cosmists*, George M. Young explained that it placed the "good of the whole community above the freedom of the individual to go his or her own independent way."[115]

Konstantin Tsiolkovsky was a friend and protégé of Fedorov. All matter consists of life, the schoolteacher believed, flowing from lower states to ever-higher orders of being. As we climb the ladder of material existence and expand ourselves into the universe, eventually reaching the pinnacle of material development, we will finally break free of our physical bodies and move into an era of gnostic-like spiritual perfection.[116]

Laboring in his homemade laboratory, Tsiolkovsky put his cosmist and scientific concepts to paper. His ideas were groundbreaking; astronautic theory, multi-stage rockets and spaceflight, artificial gravity and space-based solar energy.[117] He also designed space colonies that could sustain up to one hundred people, containers wherein the social elite would move upward while laborers worked for the collective good.[118] His technical papers formed an intellectual path that eventually culminated in the launching of Sputnik 1. Today you can see a prominent statue of Tsiolkovsky, the world's father of modern rocketry and a Soviet hero, sitting under the 107-meter tall Monument to the Conquerors of Space in Moscow. Cosmism then and today looks to space for salvation.

A cosmist contemporary of Tsiolkovsky worth noting is Vladimir Ivanovich Vernadsky. Certainly not a household name in the West, his thoughts have nevertheless soaked into our spiritual milieu and the transhuman project.

Vernadsky's importance lies in his theory of the earth evolving an intelligent biosphere. Others before him had similar ideas and Char-din's belief in the Noosphere, a consciously awakening planet, may have been formulated after attending a lecture by Vernadsky. The Russian, however, was able to press this hypothesis into a synergistic philosophy: *matter has life.*

George M. Young writes, "He was one of the first scientists to empha-size that the exchange of matter leads to a basic unity of the planet, its human inhabitants, and the cosmos."[119]

All things, according to this creed, share a unity of life. In expound-ing on Vernadsky's theory, Young contextualized it in a remarkably familiar way:

> We are, in a very deep sense, related to all on our planet – to animals, vegetables, and minerals, as well as to other human beings, and as the rational component of the biosphere, we have a responsibility, literally, to all.[120]

In light of this we are all considered global citizens, and our alle-giance *must* be to the earth—and not just as an inanimate object, but personified as thinking matter. We could call this the image of Gaia, *the living planet*, or Mother Earth. Young writes: "As inhabitants first and foremost of the cosmos and the planet, human beings owe allegiance to the biosphere more than to any other nation, ethnic entity, economic class, or system."[121]

Cosmism today is fashionable.

In my 2013 interview with Dmitry Itskov, the founder and host of the Global Future 2045 International Congress, he explained that Tsi-olkovsky was "definitely the person who inspired me." Specifically, it was Tsiolkovsky's vision of "radiant mankind and this transformation from this material body to the energetic form of a human body, it is spiritual transformation."[122]

Dmitry, whose evolutionary mission began with an experience of spiritual transformation, invoked Tsiolkovsky's name and dream during his opening speech at the GF2045. It was not the first or only time I would hear Tsiolkovsky mentioned.

The cosmist goal of unlimited transhuman freedom—escaping the bonds of earth——was also visible in the GF2045 introductory video: The planet is on the brink of global collapse, and we must either choose a new dark age or a new paradigm in human evolution; a complete technological and social revolution is therefore required if man is to survive; BCI, artificial intelligence, and simulated worlds will point the way forward; and with this revolution will come a new ideology, ethics, psychology, and culture, and even a new metaphysics. By creating manageable matter, we will move our consciousness into an *artificial carrier* and upload our personalities into a series of nearly immortal avatars. Then we can reach for the universe and ascend to the stars. The new man, unencumbered by earth and limiting matter, will then be devoted to "spiritual self-improvement."[123]

GF2045 is certainly not the only example of cosmism within the transhumanist movement. Much of the community has, in broad or specific ways, adopted some elements of the Russian philosophy. Giulio Prisco, a former analyst with the European Union Space Agency and past director of the World Transhumanist Association,[124] is a modern-day apostle. I have personally spent time with Prisco while attending a conference of the Mormon Transhumanist Association, and our avatars have crossed paths in Second Life. Prisco is devoted to the posthuman cause and advocates a new cosmism: "Science and religion, spirituality and technology, engineering and science fiction, mind and matter. Hacking religion, awakening technology."[125]

Ben Goertzel is another modern cosmist, noting that science requires faith as we look beyond the data points to seek a holistic future. Recognized for his brilliance in artificial intelligence, cognitive robotics and perceptual psychology, Goertzel holds a PhD in mathematics and is the author of *A Cosmist Manifesto*.[126] His volume bridges science and psychedelics, technology and philosophy, Carl Jung's collective unconsciousness and mathematics.

In his discussion of the "global brain," a subject he specializes in, Goertzel offers two possibilities. First, the merging of individuals within the collective intelligence, "sacrificing much of the individuality we currently associate with being human, but gaining a feeling of oneness with a greater mind." Or a system that retains a certain level of individuality without complete immersion, "connecting to the global brain could be more like using the Internet today—but an order of magnitude more pervasive."[127]

"Assuming a free society, interacting with the global brain would be optional," he wrote, "but nearly everyone would take the option, just as so many other highly convenient, inexpensive technologies have been adopted by nearly all people given the chance."[128]

He is correct. Humanity will opt in.

Transhumanism has birthed a trans-faith. In 2002 Martine Rothblatt officially launched the Terasem movement. Self-described as a *trans-religion*, Terasem, meaning "earthseed," incorporates teachings on technical immortality and mind uploading, prayers, rituals, and Kundalini yoga. It is a fusion of spirituality, visionary creativity, and a secular faith in technology.

But what does it mean to be a trans-religion?

Gabriel Rothblatt, Martine's son and a Terasem pastor, gave an explanation during the 2013 Mormon Transhumanist Association's annual conference: "Trans-religion means transcending religion. It means encompassing all religions. Essentially, secularism." Core traits of the techno-faith, Gabriel explained, are joyful immortality, unity, and diversity—"the self-fulfilling prophecy of creation."[129] Terasem's central beliefs: life is purposeful, death is optional, God is technological, and love is essential.

Martine Rothblatt argues that classical religions will eventually conform to the technological extension of humanity. Of special interest to the Terasem founder is the religious acceptance of *mindcloning*. By uploading massive amounts of personal information, creating what they call *mindfiles,* it is hoped that when some future technology comes online, these databases can be reconfigured into the "conscious prostheses of ourselves."[130] Rothblatt expects that this digital personality will then desire spirituality, and he trusts that traditional religions will recognize the copied soul of the virtual human double as valid. Equating the soul to consciousness and then extrapolating the mindclone to the function of an organ, Martine writes:

> No classical view of religion believes you transplant a soul when you transplant a heart. Nor can you kill a soul by killing the body. Hence, the mindclone continues to radiate the soul of its person whatever may have occurred to the associated body. Beautifully, this is consistent across atheism and theism....
>
> Mindcloning will be viewed as a medical technology, analogous to organ transplants, a life-extending technology embraced by even the most classically minded adherents to Judaism, Christianity, and Islam.[131]

Important to the story is Rothblatt's personal life.

Having been born as Martin in 1954, the global communication and satellite expert announced being trans-gendered at the age of forty. The following year, 1995, Rothblatt's revolutionary volume *The Apartheid of Sex* argued for gender fluidity and transgenderism as a social norm.[132] *The Apartheid of Sex* played a major role in promoting the concept of gender as a social construct; labels of male and female are acts of injustice, as sex is a complex continuum; and creative expression is negated when we constrain sex and gender to binary roles. Gender, like humanity itself, would be *trans* in the spin-cycle of social transition.

In the 2011 edition of *The Apartheid of Sex*, retitled *From Transgender to Transhuman*, Rothblatt recounts the non-binary position, adding the following in the preface:

> In a similar fashion I now see that it is also too constraining for there to be but two legal forms, human and non-human. There

can be limitless variations of forms from fully fleshed to purely software, with bodies and minds being made up of all degrees of electronic circuitry between. To be transhuman one has to be willing to accept that they have a unique personal identity, beyond flesh or software, and that this unique personal identity cannot be happily expressed as either human or not. It requires a unique, *transhuman* expression.[133]

Terasem is obscure outside of the transhumanist community. However, to posthuman thinkers and techno-utopians, this new secular religion is a beacon of enlightened progressivism.

Transhumanism has also energized the creation of a Mormon-based organization, the Mormon Transhumanist Association (MTA). Unlike Terasem, the MTA is not a religion in itself but a network to advocate for transhuman goals with an unapologetic Mormon stance: "Mormonism mandates transhumanism."[134] In other words, rapidly advancing technologies point to a fulfillment of the Mormon teaching that man is to strive for exaltation, to become as God. Although the majority of its membership is Mormon, mainly affiliated with the Church of Jesus Christ of Latter-day Saints, it is not exclusive to Mormons. Atheists, agnostics, and progressive Christians are part of its demographics. Gabriel Rothblatt has been a member of the MTA. Cosmist Giulio Prisco, an atheist transhumanist who adopted a religious position due to his MTA involvement, is a member.

Responses to the 2014 MTA *Members Survey* offer some extra details as to the composition of its body. In terms of Cultural Politics and Economic Politics, 53 percent and 31.8 percent see themselves as progressives; 75 percent are married and have never been divorced; 85.7 percent are male; and 42.2 percent have earned a post-graduate degree.[135] Interestingly, as transhumanist organizations go, the MTA has experienced relative longevity. Other groups have started and then fragmented or faded away.[136]

Formed in 2006 with fourteen people, the MTA had close to six hundred members ten years later. But numbers are always less important than narrative and reach, and in both it has demonstrated success: first in moving the conversation past secularist boundaries, and then in projecting the posthuman theme beyond its membership. Indeed, much of the recent transhuman shift from secular approaches to religious contexts have come through MTA input. And in terms of reach, the association's voice has touched a broader audience. In 2015, the journal *Theology and Science* published an essay on Mormonism and transhumanism, and articles broaching the two showed up in *HuffPost UK*, *Mail Online*, *The Daily Dot*, and *iDigital Times*. *The New Yorker* published a piece the next year, and the popular online tech-site, *New Atlas*, ran a story in early 2017. From 2010 onward the Mormon group has been mentioned in various works exploring transhumanism,[137] and the MTA itself collaborated on a book, *Parallels and Convergences: Mormon Thought and Engineering Vision*.[138]

However, the MTA's annual conference has raised its reputation as a platform to explore techno-ethics and to bolster posthuman speculations. Transhumanists and futurists like Natasha Vita-More, James Hughes, Giulio Prisco, Aubrey de Grey, Eric Steinhart, and Robin Hanson have shared the podium. Likewise, Mormon historians and authors, and MTA members—many with professional backgrounds—have presented papers and theories. And sometimes critics have been invited to speak.

I traveled to Salt Lake City in 2010 to attend the MTA conference, *Transhumanism and Spirituality*. Held on October 1 at the University of Utah, this event explored religious and technological boundaries. The MTA president outlined his *New God Argument* and Terryl Givens connected mythology and the human-divine struggle in his presentation, *Fear and Trembling at the Tower of Babel*. Others talked of humans and spiritual entities locked in a struggle for scarce resources; that all life is interconnected in the quest for spiritual perfection; and that transhumanism should be a next-gen religion intent on unifying humanity. One speaker suggested it was time for faith communities to evolve perceptions of God and spirituality, shifting theology to embrace paradox and replace dogmas with questions: "Part of evolving spirituality is paying attention to how we image God."[139] Giving a keynote, Max More contrasted atheist transhumanism against the "petulant child...cosmic sadist" of the Old Testament.

"I should tend to want to discourage talking of gods in transhumanism," More said. "I think we can probably do better than that."[140]

Evangelical Christianity was not given high marks.

I penned an article about the meeting and published it in my magazine, *Forcing Change*. Later this was circulated within the Mormon transhumanist community and elsewhere. Many months afterwards Lincoln Cannon, a cofounder of the MTA, contacted me.

Would I be willing to attend the 2013 conference to give a Christian critique of religious transhumanism?

I was surprised by the request but thankful for the opportunity, and appreciative that the organization would consider bringing a critic to the table. This was an open door and I agreed to participate.

When I arrived at the venue in Salt Lake City, Cannon—a perceptive and eloquent spokesperson for Mormonism and transhumanism—immediately shook my hand and said, "You're brave to be here." I understood the sentiment, for my evangelical beliefs and worldviews stood in contrast to those around me. But Cannon was taking a risk, too. By inviting an outside critic, he was courting a level of uncertainty, and with it, the potential for pushback from his own community. It was a gutsy move on his part and by the MTA board, who treated me with courtesy and respect.

Having only fifteen minutes to make a case, the standard time given to most of the presenters, my talk had to be concise yet meaningful. This was far from easy. For days I had struggled over my task and sleep had been elusive. My desire was to communicate in a way that was heartfelt and loving, honest, and truthful. The potential for misrepresentation and miscommunication was ever on my mind, and I wrestled with this responsibility up to the moments before my name was called.

What did I say?

The speech is publicly available,[141] but the basics are thus: An acknowledgment that Christianity is not opposed to human betterment through technology; however, it is the issue of salvation that is at the core of its tension with transhumanism. After giving a brief survey of the biblical perspective of God's relational design in Genesis 1, that all things were created by Christ Jesus as the Author of Life (John 1:3), humanity broke fellowship with God and therefore chose the path of death, and Jesus Christ entered the world of men—experiencing death but defeating this enemy in a finalizing act. I closed by focusing on Christ's crucifixion, specifically the one thief hanging by His side.

Allow me to expand on this.

In preparing for my talk, I was struck by what took place at the cross. Jesus, hanging naked and scourged and beaten, was being publicly executed between two thieves. Consider the narrative from the Gospel of Luke:

> Then one of the criminals who were hanged blasphemed Him, saying, "If You are the Christ, save Yourself and us."
>
> But the other, answering, rebuked him, saying, "Do you not even fear God, seeing you are under the same condemnation? And we indeed justly, for we receive the due reward of our deeds; but this Man has done nothing wrong." Then he said to Jesus, "Lord, remember me when You come into Your kingdom."
>
> And Jesus said to him, "Assuredly, I say to you, today you will be with Me in Paradise."[142]

Specifically, it was what was *missing* that caught my attention.

Notice that Jesus did not tell the man he would be saved by doing good works—in fact, the thief was dying because of his crimes. Good works were out of the question. Nor did He say to the thief, "Find a technical or scientific solution to your problem of death." Neither did He say to join the priesthood, or follow temple obligations, or start a charity, or go to church, or even be baptized. There was no eight-point path to enlightenment, no constructing "Heaven on Earth," no postures for liberation, no chants or spells or rituals. The thief was being executed *now* and could do *nothing practical.*

And this returns us to our present state of affairs. Like the thief, there is nothing technical to be done on our part. For us, the problem is *positional.* Just as one man's act of disobedience to God, Adam in the Garden, became representative of the human race, condemning us to death without our say in the matter, so our salvation could only come through one Man being obedient, without guilt, having the ability to overcome death—without our say in the matter.

This would require none other than the Author of Life to directly intervene. It would have to be *His perfecting act* given as a gift. The Apostle Paul put it this way: "For by grace you have been saved through faith, and that not of yourselves; it is the gift of God, not of works, lest anyone should boast." The implication is clear: plusses are not going to work. It cannot be Jesus Christ *plus* technology *plus* politics *plus* religious rites and obligations, plus, plus, plus. The plusses are never ending and never good enough.

In John 14:6 we read the words of Jesus: "I am the way, the truth, and the life. No one comes to the Father except through Me." There may be some temporary good in the plusses, but they are incapable of saving. The real question is this: Will we trust in Jesus Christ for our salvation?

Belief and trust in Jesus Christ was all the robber had left, and that the thief recognized his own sin demonstrates a contrite heart.[143] Belief opens the door to relationship, repentance to forgiveness, trust to expectant hope. Jesus Christ said to him, "Today you will be with Me in Paradise." For it would be Christ, by the power in Himself, in God the Father and the Holy Spirit—the Deity-in-trinity[144]—through which death would be conquered three days later. As the Apostle Paul said in his letter to the early Christians, Jesus Christ is the "firstborn from the dead, that in all things He may have the preeminence."[145]

The thief's soul was safe in the *perfect position* of the Redeemer.

For those who, like the thief, admit our inability to save ourselves—who believe and repent, entering a relationship of trust with the Author of Life—a new man is born.

This is not the new man of Marx or Lenin. This is not the Supreme Being of Comte or the *Übermensch* of Nietzsche or the master race of Nazi Germany. Neither is this the perfected man of theosophy or the "good man made better" of Freemasonry or Chardin's Ultraman. It is not exaltation through the rituals and obligations of Mormonism. Nor is it the neo-human dream of eugenicists or the posthuman vision of transhumanists.

No, the New Man in Christ—the person whose salvation is placed in the *already resurrected* Messiah—is *new* in a different way. Redeemed in the position of Christ, the person is now spiritually aware and awake. The ancient letter to the church in Ephesus illustrates what this looks like: to no longer walk in the futility of a darkened understanding, a person alienated from God, living a life of self-gratification and greed. Instead, walking in the truth of Jesus Christ, to discard our former conduct, "the old man which grows corrupt" and to renew the spirit of our mind, "that you put on the new man which was created according to God, in true righteousness and holiness."[146]

The character of the New Man is to be established on the image of God, graced with tender mercies, kindness and humility, expressing power with reservation and gentleness—the meaning of meekness—to be longsuffering, "bearing with one another, and forgiving one another." The New Man is to put on love, "which is the bond of perfection" and to "let the peace of God rule in your hearts." We are to be thankful to God, acknowledging Jesus Christ in our words and actions.[147]

We are saved through faith, but after salvation the New Man is to exemplify this in what we say and do. The new life is active, not passive. Faith without works is a dead religion.[148]

I will be honest. While my salvation is in the finished work of the Messiah, the old man in me battles with the New Man I am supposed to be. A day is coming however, when those in Christ will be raised by Him in spirit and body: *Resurrection to life.*[149]

Vacation Death

What the Bible says about trusting Christ for perfection and a coming resurrection is dismissed by atheist transhumanists, even as they strive for perfection and resurrection. Religious transhumanists, on the other hand, tend to add plusses to what Christ accomplished. Of course, if what the Creator of Life did was insufficient, then mankind as a created being is even more helpless. The problem cannot be the solution.

Scientifically escaping personal extinction and reanimating the dead are inescapable themes in transhumanism. It was manifest at GF2045.

Dmitry Itskov put this into a utopian framework, telling me that racial and national tensions will cease, and war will become impossible "because there will be no death."[150] By 2045, Itskov and others contest, our minds "will become substrate-independent." In other words, our transferable and immortal consciousness will be able to choose its own body types, shape shifting nanorobotic forms or "body holograms featuring controllable matter."[151] Peter Diamandis projected his idea of an "immortal planetary meta-intelligence," writing in the congress program: "We will no longer have to die a physical death, enabling who we are…to continue for a far longer time."[152]

While at GF2045, I had the opportunity to interview Natasha Vita-More, the wife of Max More and a transhuman designer passionate about "whole body prosthetics." She speculated what death might look like in a *post-terminal* existence:

Death ought not to just have one express definition…we will be redefining death the more and more we develop the sciences and technologies to intervene with death. So, we have to look at death from various perspectives. You may want vacation death.

You may want partial death. You may want to be put into cryonic suspension or another type of preservation unit…and then
come back and revisit life again. So there are all sorts of alterna-
tives to this finality of death."[153]

Examples of the transhuman dream of death-suspension and resurrection abound.

Before transhumanism became the intellectual movement it is now,
American writer Alan Harrington offered a survey of immortality. His
perspective demonstrated the anti-God sentiment of the 1960s cultural
shakeup:

Death is an imposition on the human race, and no longer
acceptable. Men and women have all but lost their ability to
accommodate themselves to personal extinction: they must now
proceed physically to overcome it. In short, to kill death.…

Our survival without the God we once knew comes down
to a race against time. The suspicion or conviction that "God
is dead" has lately struck home not merely to a few hundred
thousand freethinkers but to masses of the unprepared. Ancient
orthodoxies may linger, but the content of worship has begun
to collapse. This is what makes our situation urgent: around the
world people are becoming increasingly less inclined to pray to
a force that kills them.[154]

Years later, mathematical physicist Frank Tipler postulated a naturalist theory of eternal life in his book, *The Physics of Immortality*.

According to the Tulane University professor, the Christian belief
in a spiritual being called God and the non-material idea of the soul are
faulty but not beyond redemption. God, heaven, and life evermore can
be manifested through quantum mechanics in a coming Omega Point.
The monotheistic faiths, he argued, are expressions of mathematical
constructs: "The key concepts of the Judeo-Christian-Islamic tradition

are now scientific concepts. From the physics point of view, theology is nothing but physical cosmology based on the assumption that life as a whole is immortal."[155]

Tipler's eternal life theory, riding the cosmist theme of *material resurrecting material*, understandably received pushback from his academic peers. However, what was once ridiculed as scientific absurdity is chic philosophy—the allure of *mystical materialism*.

But something else needs to be considered: Just because techno-elites are serious about scientific immortality does not mean their naturalist dream is within reach. At GF2045, one transhuman philosopher told me that he had been striving for resurrection most of his adult life. Quieting his voice in sober reflection, he acknowledged that for almost thirty years this had been his primary pursuit, yet, with all the technical advances of the past three decades, he was not one day closer to immortality. The irony did not escape me. I too believe in a coming resurrection and look forward to that glorious day, but our faiths are placed on different foundations: one on the shifting sands of silicon, the other on the Rock of Ages; one on the endless toil of fallible hands, the other on the Author of Life—*who already proved Himself by walking out of the tomb*.

Breaking the death barrier is the hope of famed innovator Ray Kurzweil. In his bestsellers, *The Age of Spiritual Machines* and *The Singularity Is Near*, the technologist proposed that our memory information—our *mind files*—could be the keys to immortality. Correlating the human body to computer hardware, he advocated for backing up our personalities:

> Up until now, our mortality was tied to the longevity of our hardware. When the hardware crashed, that was it.... As we cross the divide to instantiate ourselves into our computational technology, our identity will be based on our evolving mind file. *We will become software, not hardware....*
>
> As software, our mortality will no longer be dependent on the survival of the computing circuitry. There will still be

hardware and bodies, but the essence of our identity will switch to the permanence of our software. Just as, today, we don't throw our files away when we change personal computers—we transfer them, at least the ones we want to keep. So, too, we won't throw our mind file away when we periodically port ourselves to the latest, even more capable, "personal" computer....

Our immortality will be a matter of being sufficiently careful to make frequent backups.[156]

Kurzweil is equating *information with life*.

"Ultimately software-based humans will be vastly extended beyond the severe limitations of humans as we know them today," he writes.[157] And in evolving away from personal extinction, he muses, we will no longer need to rationalize our demise. Death will be defeated by data.

The Terasem movement embraces the mind file concept by encouraging its members to partake in *mindcloning*, the creation of a digital double that will supposedly live beyond biological death through virtual reality. Mind-file dossiers are constructed by having Terasem members submit as much personal information as possible, as often as possible: pictures and other media, documents, and journal entries detailing one's experiences and emotions. Terasem can also bank samples of your DNA and in the unknown future when science supposedly catches up with the dream, your mind-file memories will be uploaded to a reconstructed body.

What is important to keep in mind is this: Everything being described is driven by a determined worldview. Consider the words of Giulio Prisco:

We will develop spacetime engineering and scientific "future magic" much beyond our current understanding and imagination. Spacetime engineering and future magic will permit achieving, by scientific means, most of the promises of religions—and many amazing things that no human religion ever dreamed. Eventually we will be able to resurrect the dead by "copying them to the future."[158]

Zoltan Istvan offered an imaginative scenario during his 2015 Terasem Island talk, suggesting that the Christian "heaven" could be constructed in virtual reality. Or through "quantum archeology," a *time-magic technology*, we might "come up with a way to resurrect every single human being that ever lived, at any time in history, and give them a chance to live again, if they want—or give them a chance to just be dead." This "Jesus Singularity," he believed, would fulfill Christian prophecy.[159]

Aging and death: If these two fundamental problems could be overcome, then tensions between people and conflict between nations could be resolved. Such thinking lies behind RAAD Fest, a project of the Coalition for Radical Life Extension. Located in beautiful San Diego, California, RAAD—Revolution Against Aging and Death—is an immersive and collaborative transhuman gathering designed as the "Woodstock of radical life extension."[160] Part conference, part festival, RAAD brings together leading thinkers, tech innovators and scientists, futurists, artists, and musicians in celebration of immortality. Together through the power of our knowledge, we can live forever, together we reach for something more.

It looks like we are racing to a Second Genesis: Not a re-creation but a *re-fall*.

In reviewing the contemporary transhuman theme of immortality, I am reminded of Alan Harrington's earlier prediction of a transitional generation. After examining the state of 1960s medical knowledge and the promise of cryonics, he forecast a coming period that would lie between the dying society and the rise of immortalists:

Members of the transition generation will almost surely not live to experience the immortal state. Knowing this, we will have to psych ourselves, like athletes, into a superior performance.

We can begin with a self-congratulatory religion rather than a humbling one, spreading abroad a new faith, honoring our race instead of punishing it for an imaginary primal crime. Through our efforts we honor the human species by helping to turn it into the divine species.[161]

Some contemporary transhumanists might see themselves as the transitional generation, what Harrington described as "the heroes and heroines of the evolutionary process."[162]

For others, hope and trust is centered on finding a technological solution to death before their passing. From either perspective, however, to be truly posthuman requires mastery over mortality—to enter the realm of divinity, *to be as God*.

Being and Becoming

Transhumanism is the quest to move from *being* to *becoming*: from being human with all of its mortal flaws to becoming divine-like in knowledge and capacity. God-talk, regardless of how individual transhumanists frame deity, is rife within the community. It runs three ways: We become gods, or our machines become gods, or by our techno-evolution we awaken the universe to its god-state. All three speak to one boast: *Man is God*.

The Mormon view often communicates the first position. In *Parallels and Convergences: Mormon Thought and Engineering Vision,* cosponsored by the Mormon Transhumanist Association, we read the following:

The end point of engineering knowledge may be divine knowledge. Mormon theology permits us to think of God and humans as collaborators in bringing to pass the immortality and eternal life of man. Engineers may be preparing the way for humans to act more like gods in managing the world.[163]

Lincoln Cannon's opening speech during the 2013 MTA conference followed a similar pattern:

> The Mormon Transhumanist Association stands for the propo-
> sition that we should learn to become Gods, and not just any
> kind of God, not the God that would raise itself above others,
> but rather the God that would raise each other together. We
> should learn to become Christs, saviors for each other, consolers
> and healers, as exemplified and invited by Jesus.[164]

Obviously, I fundamentally disagree with this assertion.

However, Lincoln's attitude is much different than Richard Seed's
approach, and I compare the two only to show the range of thinking
found within the broad community. While Lincoln hopes for a mutu-
ally beneficial god-state, Seed, the physicist who made national news for
supporting human cloning in the late 1990s, allowed hostility to show
in an interview for the documentary, *Technocalypse*:

> We are going to become Gods. Period. If you don't like it, get
> off. You don't have to contribute; you don't have to participate.
> But if you're going to interfere with me becoming God, we're
> going to have big trouble. Then we'll have warfare.[165]

Seed went on to say, "The only way to prevent me is to kill me; and
you kill me, I'll kill you."[166]

It would be easy to dismiss Seed's rhetoric as hubristic nonsense,
especially in that before the national spotlight turned on him for his
cloning advocacy, he was a little-known figure. However, Seed's words
play into a contextual reality: The history of man "playing God" is one
of deepest subjugation and destruction. We proclaim the good but are
unable to break from that which is evil; we are incapable of resisting the
allure to become a god *made in our fallen image*.

When Kevin Kelly, then editor of *Wired* magazine, wrote his 1994 book *Out of Control,* he dabbled in god-talk. Considering the direction of virtual reality and the possibilities of a techno-hive, Kelly recognized the appeal of human deity through technology:

> Stripped of all secondary motives, all addictions are one: to make a world of our own. I can't imagine anything more addictive than being a god…. Godhood is irresistible.[167]

Could our technical creations be set loose to discover their own sovereignty? Might our innovations make some future attempt at *becoming*? Kelly toyed with these underlying ideas:

> As we unleash living forces into our created machines, we lose control of them. They acquire wilderness and some of the surprises that the wild entails. This, then, is the dilemma all gods must accept: that they can no longer be completely sovereign over their first creations.
>
> The world of the made will soon be like the world of the born: autonomous, adaptable, and creative but, consequently, out of our control. I think that's a great bargain.[168]

John Glad pondered something similar: "Will the computer have a God? Will God have a God?"[169]

Comparing "carbon-based technology"—human beings—to computational technology, it was clear to Glad that the human brain was too restricted by its size and speed and time needed for learning. Limitless machines would therefore win the day:

> Any effort to improve the human brain is targeted at an instrument which is inherently limited in its capacity. The machine brain, on the other hand, will be something like God.[170]

Ben Goertzel wrestled with ideas of *evolving mind* and *building god*:

Whether or not transhuman minds now exist in the universe, or have ever existed in the universe in the past, current evidence suggests it will be possible to create them—in effect to build "gods." As well as building gods, it may be possible to become "gods." But this raises deep questions regarding how much, or how fast, a human mind can evolve without losing its fundamental sense of humanity or its individual identity.[171]

Ray Kurzweil offered a "God" viewpoint through his Singularity model, the reaching of a universal divine-state through the expansion of our technological stepchildren:

We can consider God to be the universe…. The universe is not conscious—yet. But it will be. Strictly speaking, we should say that very little of it is conscious today. But that will change and soon. I expect that the universe will become sublimely intelligent and will wake up.[172]

Mark Pesce asserts the immortalist divinity of man:

Once the genome was transcribed, once we knew what made us human, we had, in that moment, passed into the Transhuman.

Knowing our codes, we can recreate them in our so-called synthetic worlds of ones and zeroes: artificial life.

Now that we have discovered the multiverse—where nothing is true, and everything permissible—we will reach into the improbable, resequence ourselves into a new being, debugging the natural state, translating ourselves into supernatural, incorruptible, eternals.

There is no god but Man.[173]

Upon the Tower of Technology, we erect idols to ourselves.
At the end of *The Phenomenon of Science*, Soviet cybernetic vision-

ary and mathematician Valentin Turchin pondered the interconnecting paths of evolutionary futures. Looking at humanity from outside itself to gain the "scientific knowledge of reality," he closed with these chilling words:

> We have constructed a beautiful and majestic edifice of science. Its fine-laced linguistic constructions soar high into the sky. But direct your gaze to the space between the pillars, arches, and floors, beyond them, off into the void. Look more carefully, and there in the distance, in the black depth, you will see someone's green eyes staring. It is the *Secret*, looking at you.[174]

The Future Is Past

Cybernetic medicine, brain reverse engineering, neural-dust-enabled BCI, gemenoids and telenoids and robot heads; GF2045 was a showcase of the *already* and the theoretical.

But not everything worked out.

"Dmitroid," the android head of Dmitry Itskov—a robotic copy—failed to materialize. Designed by the renowned robotics inventor David Hanson, Dmitroid was supposed to be introduced on the first day of the congress, but technical glitches kept the android from its public debut. For many participants this was a disappointing turn of events, as Hanson's robots are world-known for their lifelikeness.

But it was not the technologies that caught my attention; it was the politics and religion—and the reiteration of crisis to promote the transhuman cause.

To this end, the GF2045 program outlined the need for a new development strategy to address global challenges, and Itskov spoke of bringing lasting world peace. Evolutionary transhumanism would end competition and the struggle for survival, thus the *new human* could bathe in spiritual enlightenment.

Dr. James Martin,[175] the man who literally wrote the book on computers and networking, outlined a series of burgeoning world crises: overwhelming climate change, an out-of-control human population, resource scarcity, the destruction of agricultural capacity, and massive economic imbalances. All of this, he explained, will require a total transformation of human nature and the reshaping of our future before it arrives. Coal-fired power plants would thus have to be shut down and the world's number of domestic cattle lowered to 200,000 from roughly 1.5 billion. Solar-powered "climate change cities," he said, would sprout in the far northern and southern regions of the globe.[176] He noted, with some accuracy I believe, that Russia and China may be better positioned to survive world upheavals. And he predicted the necessity of an international technocracy to manage global society by 2045.

Transhumanism and technocracy; one is a nail and the other a hammer.

One goal of GF2045 was to compel the United Nations to take a leadership role in future human evolution. Itskov wrote in the program book: "I believe that the new evolutionary strategy should be considered at the level of large public and transnational organizations and government leaders."[177] An open letter was sent to then UN Secretary-General Ban Ki-moon, beseeching his support and calling on the United Nations to promote a "strategy for the transition to neo-humanity."

Ban Ki-moon never showed at the congress, but the petition to consider a transhuman politic demonstrated a growing political awareness.

Religion, too, was expressed.

From the opening speech to the last day's interfaith panel, spirituality and religion were intertwined within the techno glitter. Cosmism was on display, it was true, but Itskov had opened the door for other spiritual paths. Tibetan Buddhists and Hindu Monks and spiritual sojourners joined in discussing evolutionary transformation.

Quantum activist Amit Goswami argued for the *ground of all being*, the oneness of science, spirituality, and psychology. Transcendent consciousness *is* reality, he explained, calling for a reconfiguration of human nature and the evolution into a quantum society. A one-world soul is encompassing mind and body, matter and consciousness—*we are all one.*[178]

Swami Vishnudevananda Giri, the founder of a Russian yogi school, promoted an evolutionary model beyond the technological. *Transcendent transhumanism* is the goal; science united to spirituality, and the intentional recognition of our God-like nature. He reminded us that the first transhumanists were yogis and spiritual masters, and that our task is to form a God-like civilization and become immortal God-men—controlling thought and time, and possessing immortal bodies located in multiple realities: "Fantasy will become reality."[179]

Hindu yoga master, Mahayogi "Pilot" Baba, a former wing commander in the Royal Indian Air Force, said something similar: "You are the light, you are the truth, you are the beginning, you are the end."[180]

I recognized where Mahayogi Baba pulled those words. He had borrowed the sayings of Jesus Christ, creating a montage of self-serving spirituality in the place of the Messiah's exclusivity. Baba had combined John 14:6 and Revelation 22:13:

> Jesus said to him, "I am the way, the truth, and the life. No one comes to the Father except through Me." (John 14:6)

> I am the Alpha and the Omega, the Beginning and the End, the First and the Last. (Revelation 22:13)

The words of Jesus Christ in Revelation 22 parallel God's powerful declaration as found in Isaiah 44. Indeed, Jesus Christ is the Redeemer, "the Lord of hosts," declaring Himself as "the First and the Last":

> Thus says the Lord, the King of Israel, and his Redeemer, the Lord of hosts: I am the First and I am the Last; Besides Me there is no God…. Is there a God besides Me? Indeed there is no other Rock; I know not one. (Isaiah 44:6,8b)

The writer of Ecclesiastes famously said there is "nothing new under the Sun."[181] And he was right. "*You will not surely die…you will be like God…knowing good and evil.*"[182]

Social critic Neil Postman is spot-on: "'Thou shalt have no other gods before me' applies as well to a technological divinity as any other."[183]

Yearning for technical oneness through our "machines of loving grace," transhumanism is man becoming what we worship.

It is just another *game of gods*.

Transhumanism
Artificial Salvation, Virtual Heaven, and Synthetic Immortality

By Milieu Member Paul McGuire

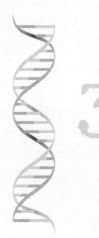

In his intellectually robust article, "Secularizing Demons: Fundamentalist Navigations in Religion and Secularity," S. Jonathon O'Donnell lays down the gauntlet on what he categorizes as a loose-knit group of post-evangelical intellectuals and best-selling authors led by author, theologian, and founder of SkyWatch TV, Tom Horn, whom O'Donnell suggests blends "post-apocalyptic thinking, transhumanism, and the Bible's discussion of the reality of a personal evil led by Lucifer."

O'Donnell calls this group "The Milieu," which is a name borrowed from a category of French organized criminal groups that once operated from Paris, France. In a very real sense, O'Donnell's use of the term "The Milieu" is an excellent choice for Horn, who, by implication, is the intellectual and theological "godfather" of this eclectic group of authors, speakers, video producers, television and radio broadcasters, and social media entrepreneurs who collectively reach over one hundred million globally.

As the author of this chapter, I would like to establish the fact that I was raised in an atheistic, intellectual, and humanistic home in New York City, where, beginning as a young child, I was exposed to endless discussion and debates among my parents' humanistic friends who were artists, writers, scientists, and intellectuals.

I was taught at an early age that Christianity was a religion for ignorant and superstitious people who were "anti-joy," "anti-life," and "anti-sex." The two books I read in third grade and thoroughly understood were *Brave New World* by Aldous Huxley and *1984* by George Orwell. These became two of the most influential books in my life. At about the age of ten, I decided that I wanted to become either a nuclear physicist (after I read a book on Albert Einstein), or an oceanographer (after I read the writings of Jacques Cousteau, a religiously tolerant, but bold, humanist).

I also read the works of Erich Fromm, Dr. Milton Erickson, B. F. Skinner, Frederic Nietzsche, Jean Paul Sartre, Dr. John von Neumann, Dr. John C. Lilly, Abraham Maslow, R. D. Laing, Sigmund Freud, and others. I was a huge reader of science fiction from authors like Isaac Asimov and Robert Heinlein, and became self-educated in neurological science, psychology, consciousness, biology, physics, and other disciplines.

I attended the University of Missouri, where I majored in filmmaking and a new field of psychology called "altered states of consciousness." All through this time, I was militantly anti-Christian, as my observations of American Christianity led me to believe it was a religion of fools, and I saw no reality in what Christians claimed to believe versus how they acted and what their one-dimensional culture produced. I very much agreed with the essay, "Why I Am Not a Christian" by Bertrand Russell based on a talk he gave on March 6, 1927, at the National Secular Society.

A Committed Young Transhumanist

Without even being aware of the term, I was a young but committed transhumanist. Although the childhood experiments I conducted certainly would not have passed the rigors of a truly scientific method, they were an awkward attempt to prove the beginning of a childhood belief in transhumanism.

First, while I was in junior high school, I decided to conduct some amateur cryonics experiments involving freezing plants. I had read articles about cryonics, cryogenics, and about how certain very wealthy people were freezing their bodies in highly secure underground facilities so they could be brought back to life in the future when the technology had been perfected to, in a sense, "resurrect" them.

In high school, after reading Huxley's *Heaven and Hell* and *The Doors of Perception*, I embarked on an amateur scientific experiment with a fellow honor student whose father was a medical doctor. The two of us replicated Huxley's experiments in consciousness, which involved experimenting with the psychedelic drugs mescaline and LSD. Although the term "transhumanism" was not part of the lexicon at this time, the purpose of my experiment was to see if psychotropic drugs could enhance human intelligence and perception, and raise our consciousness to a superior level based on the works of Huxley and Harvard University Professor Dr. Timothy Leary. Leary developed a transhumanist and future philosophy that he summed up in the phrase "S.M.I2.L.E.,"— "Space Migration, Increased Intelligence, Life Extension."

In *Transhumanist Plus*, author Michael Garfield wrote an article titled, "The Psychedelic Transhumanists." He studied the various kinds of transhumanism visionaries such as computer scientists, genetic engineers, and what he terms "The Psychedelic Transhumanists," like the late Dr. Leary, who predicted the convergence of psychedelics, technology, and pharmacology, along with enhancing human intelligence through drugs and computer technology.[184]

Another person I began studying in high school was Dr. John C. Lilly, an American physician, neuroscientist, psychoanalyst, "psychonaut," and the author of such books as *Programming and Metaprogramming in the Human Biocomputer: Theory and Experiments and The Center of the Cyclone*.

When I was attending the University of Missouri and majoring in "altered states of consciousness" and filmmaking, I attempted to duplicate Lilly's experiments by constructing my own sensory-deprivation chamber in a secure, soundproof room that shut out all sensory input like light and sound, where I would float in body temperature water and attempt to achieve what Lilly called a satori state of consciousness to develop super-intelligence.

I remember reading that, in 1934, Lilly had also read Huxley's *Brave New World*, and he observed that Huxley had written about the use of drugs and the links between chemical processes and subjective experiences of the mind. Lilly's maxim was: "In the province of the mind what one believes to be true, either is true or becomes true within certain limits. These limits are to be found experimentally and experientially. When so found these limits turn out to be further beliefs to be transcended. In the province of the mind there are no limits."

In *The Human Biocomputer* in 1974, Lilly wrote: "For the first time I began to consider that God really existed in me and that there is a guiding intelligence in the universe."

The 1980 movie, *Altered States,* starring William Hurt was based on Lilly's work, as was the 1973 movie *The Day of the Dolphin* starring George C. Scott. When Lilly referred to "God," he wasn't necessarily referring to the Christian God, but rather to his contact with highly intelligent entities. Lilly believed that both the Christian God and these entities were beliefs to be transcended, and that beyond that there was some kind of god in the universe, not yet defined.

Dr. Leary, Ken Kesey, and Huxley all referred to some type of mystical force or higher consciousness when they used psychedelic drugs. Ironically, their books and practices opened the doors of perception in my own mind about the reality of some kind of god, which at that time would have involved higher consciousness. The psychedelic transhumanists and many leaders in the humanist movement were no longer traditional humanists in the sense that they entered a kind of religious mysticism.

The Search for Truth

In my search for truth about life's existence years earlier, at the age of 15, I had joined the "peace, love, sex, drugs, and rock 'n' roll" counterculture of the early 1970s. However, I slowly began to perceive inconsistencies in humanistic revolution regarding the love and compassion it claimed to have for mankind versus the reality.

I also became increasingly aware of the lack of scientific, empirical evidence regarding many things that humanism claimed to be true. On one hand, I found Christianity repulsively anti-intellectual, but on the other, I noticed an increasing rejection of the scientific method and empirical evidence and a lack of reason among certain humanist beliefs.

For example, I became part of the counterculture of the 1970s in the belief that mankind and the world could evolve into a one-world planetary government through scientific higher consciousness in which things like love, compassion, and peace would prevail. I thought mankind through science and human enhancement of intelligence through drugs and technology would enable us to transcend wars caused by nationalism and the independent nation-state, and that through science we could end poverty, hunger, and disease.

However, through my numerous interactions with leaders of the counterculture, I saw intense selfishness, power struggles, and a lack of love and compassion. During this time, I demonstrated with radical activist Abbie Hoffman and spoke to Kesey, author of *One Flew over the Cuckoo's Nest,* in Central Park. Kesey volunteered to take LSD through a government program run through the Stanford Research Institute.

I was with Dr. Leary, the Harvard University professor who promoted LSD, in the East Village. As I will explain later, at this time in my life I didn't understand that the CIA was working with British intelligence and was financing people like Kesey, Dr. Leary, Dr. Lilly, and others. This is why both Kesey and Dr. Leary were able to get their hands on hundreds of thousands of doses of LSD to distribute freely through Kesey's "Electric Kool-Aid Acid Test" and Grateful Dead concerts. Despite Dr. Leary's involvement in distributing unlimited doses of LSD, neither of these men ever got serious jail time.

Ominously, LSD was the drug used by Nazi scientists who experimented with people being programmed through mind control to become Manchurian Candidates, assassins, and beta-programmed sex slaves in the CIA's MKULTRA program. The question that should be asked is why a handful of men like Huxley, Dr. Leary, Kesey, and others collectively distributed millions of doses of LSD and turned on the counterculture to sex, drugs, and rock 'n' roll. The other primary components of mind control revolve around pain or intense shock, drugs, and hypnosis—often produced through repetitive music with subliminal and hypnotic lyrics.

When you combine the shock and pain of millions of Americans seeing their friends come home dead from the Vietnam War, the television footage of bombings, napalm, and death, along with President John F. Kennedy getting his head blown off while in an open convertible with his wife, Jackie, in Texas, and the bloody Charles Manson killings, you have injected massive psychic pain and shock into the American consciousness. Then consider the most powerful mind-control drug known to mankind like LSD, and the hypnotic and subliminal lyrics of the Beatles, Rolling Stones, Led Zeppelin, the Doors, and other rock 'n' roll groups, and you'll get a glimpse of what we're confronting today.

Escape from Reason

At the University of Missouri, I read a book called *Escape from Reason* by Dr. Francis Schaeffer.

Dr. Schaeffer was a Christian theologian and philosopher unlike any Christian author or faith leader of his time in that he was not anti-intellectual and his books spoke about art, culture, science, movies, philosophy, humanism, genetics, the Enlightenment, and the Age of

Reason. Schaeffer, who lived in the mountains of Switzerland, would often walk up the road to the chalet where Dr. Leary lived at the time, and the two became friends and had intense conversations. Dr. Schaeffer was a genius, and he would write in his books about the irrationality of humanism on an intellectual basis. His *Escape from Reason* was the most powerful book on culture and philosophy I had ever read. It was light years ahead of the works of Fromm and Herbert Marcuse. Dr. Schaeffer opened the door in my mind that there was an intelligent possibility of the existence of what he termed "The Infinite Personal Living God of the Universe."

At this point, I noticed that humanism, left to its own devices, always seemed to devolve from idealist and utopian ideas into ugly, cruel, barbaric, dehumanizing, and totalitarian states. I would venture to say that most humanists and transhumanists are ethical and compassionate people, but that their idealism blinds them to man's fallen nature. This is the denial of empirical scientific evidence about the history of mankind and an escape into a non-scientific Romanticism.

During this time, I didn't know that the American counterculture was a product of mass social engineering orchestrated by Fabian socialists and humanists like Wells, Aldous Huxley, Julian Huxley, the Tavistock Institute, British and American intelligence agencies, and Bertrand Russell, who organized the peace movement. Incorporating the principles of Aldous Huxley's "scientific dictatorship," they first used the British invasion of rock 'n' roll groups like the Beatles, the Rolling Stones, the Who, Led Zeppelin, Pink Floyd, and others whose albums were played repetitively via records and radio stations and contained overt programming messages on the conscious and subconscious levels. This mass mind-control operation was also used through American rock groups such as the Doors, Jimi Hendrix, the Grateful Dead, and others.

Wells, the great science-fiction writer of novels like *The Invisible Man* and *The War of the Worlds*, was the head of British intelligence during World War I. He personally mentored Aldous and Julian Huxley, who became high-ranking officers in British intelligence and introduced them to satanist Aleister Crowley, who was an asset of British intelligence and assisted British Prime Minister Winston Churchill in defeating Adolf Hitler during World War II.

Publicly, Wells, the Huxleys, and others pretended to be purely secular humanists, but privately, they had adopted some of the doctrines of Freemasonry, such as the essentially transhumanist doctrine of man evolving into or becoming gods. The influence of Crowley's occult beliefs on culture is evidenced by the fact that his picture is on the Beatles' *Sgt. Pepper's Lonely Hearts Club Band* album cover, and his name and teachings continually pop up in culture and the music industry today.

The "Scientific Dictatorship" and Mass Mind Control

What most people don't realize is that Aldous Huxley's "scientific dictatorship" is based on using mind control to rule the masses without them being aware of it, and to program them to love their slavery. Globally, we have been living under the soft totalitarianism of humanism and increasingly a "scientific dictatorship" in which the globalists elite secretly rule.

The goal of the artificially created counterculture was to take down Christianity, marriage, traditional moral values, nationalism, patriotism, and capitalism, which all stood in the way of establishing a one-world socialist state. It's not an accident that Aldous Huxley moved to Los Angeles in 1936 to promote the use of the drug mescaline among the Hollywood elite. Huxley was a key "change agent" who came to America to spread the use of LSD and transform mass control.

Zbigniew Brzezinski, one of the most powerful members of the globalists elite who masqueraded as a compassionate humanist until his death several years ago, clearly outlined the elite's plan for mankind as far back as 1971.

"The technetronic era involves the gradual appearance of a more controlled society," Brzezi ski wrote in his book, *Between Two Ages: America's Role in the Technetronic Era*. "Such a society would be dominated by an elite, unrestrained by traditional values. Soon it will be possible to assert almost continuous surveillance over every citizen and maintain up-to-date complete files containing even the most personal information about the citizen. These files will be subject to instantaneous retrieval by the authorities."

Finally, Brzezinski reminds the elite that this same technology allows them to redirect and control society in ways never before imagined, which will allow the elite to rule the masses through a global "scientific dictatorship." Here we see how the utopian vision of humanism, or creating a man-made paradise on earth, always creates a dynamic where the end justifies the means.

"The potential for controlling the masses has never been so great, as science unleashes the power of genetics, biometrics, surveillance, and new forms of modern eugenics (transhumanism); implemented by a scientific elite equipped with systems of psycho-social control (the use of psychology in controlling the masses)," Brzezinski wrote.

Among the various monsters who call openly for a "scientific dictatorship" was Dr. Jose Delgado, the director of neuropsychiatry at Yale Medical School who experimented on humans and advocated mind control. "We need a program of psychosurgery for political control of our society," Delgado said. "The purpose is physical control of the mind. Everyone who deviates from the given norm can be surgically mutilated.... The individual may think that the most important reality is his own existence, but this is only his personal point of view. This lacks historical perspective. Man does not have the right to develop his own mind. This kind of liberal orientation has great appeal. We must electronically control the brain. Someday armies and generals will be controlled by electric stimulation of the brain."[185]

Dr. Delgado, both a humanist and transhumanist, believes that as a member of the scientific elite, they must force the masses to submit to psychosurgery for political domination of a society. Dr. Delgado made other statements like the one above, plus he spent a lifetime developing the technology to do it.

But read the glowing "Tribute to Jose Delgado, Legendary and Slightly Scary Pioneer of Mind Control" article in the *Scientific American* magazine, which carefully omits the full truth about a human monster. For *Scientific American* to write this is like writing a tribute to Hitler and all the Nazi scientists experimenting on concentration camp victims. This is an example of the irrationality of humanism and transhumanism—that unspeakable acts against humanity are somehow justified in the name of science.

"Once among the world's most acclaimed scientists, Jose Manuel Rodriguez Delgado has become an urban legend, whose career is shrouded in misinformation," John Hogan wrote in *Scientific American*. "Delgado pioneered that most unnerving of technologies, the brain chip, which manipulates the mind by electrically stimulating neural tissue with implanted electrodes."[186]

Scientific American writes a tribute to one of the most anti-human and evil neuroscientists who justified all he did in the name of a psychocivilized society. The publication claims that the criticism of Delgado is based on a conspiracy theory, so are we to assume that *Scientific American* rejected the scientific method by rejecting the numerous documented quotes by Delgado?

A Clockwork Orange World

The "scientific dictatorship" that is quickly accelerating all around us began to really take off during the counterculture of the early 1970s, when our beliefs, culture, and values went through a radical social engineering process. Very few thinkers at the time realized this and were willing to speak out about it. The great filmmaker, Stanley Kubrick, did so through his films such as *A Clockwork Orange*, *The Shining*, and *Eyes Wide Shut*.

In his 1928 book, *Propaganda,* Edward Bernays, the father of modern advertising and public relations, wrote:

If we understand the mechanisms and motives of the group mind, it is now possible to control and regiment the masses according to our will (the scientific elite) without their knowing it. Those who manipulate this unseen mechanism of society constitute an invisible government which is the true ruling power of our country....
In almost every act of our daily lives, whether in the sphere of politics or business, in our social conduct or our ethical thinking, we are dominated by the relatively small number of persons...who understand the mental processes and social patterns of the masses. It is they who pull the wires which control the public mind.

The philosophy that allowed all this to happen is humanism and transhumanism, with the belief that a genetically superior elite has the Darwinian right to use science and technology to control and enslave the masses.

In addition, the humanist belief that it's possible to create paradise, heaven on earth, or a utopia through science and technology is the foundation of this philosophy.

The Age of Enlightenment and the Age of Reason began with Sir Francis Bacon's *Novum Organum* in 1620 and ended with the French Revolution in 1789. Early philosophers like Immanuel Kant (1724–1804), who originally promoted the Enlightenment, ended up criticizing it because it had succumbed to irrational Romanticism of philosophers like Jacques Rousseau, which was an early form of New Age mysticism, somewhat like that of the "hippie" counterculture movement of the 1960s that was a product of social engineering.

Simultaneously with the scientific revolution launched by Aldous Huxley, Dr. Gregory Bateson, and others, the Macy Conferences in Manhattan gathered a consortium of neurological scientists, mind-control experts at intelligence agencies, the advent of cybernetics, the work of Jon von Neumann, B. F. Skinner and B. F. Skinner's *Walden Two*, which promoted programming man and the masses like a computer through "behavior modification."

Isaac Asimov, author of the *I, Robot* science-fiction series, R. Buckminster Fuller, inventor of the Geodesic Dome, and others functioning as "gods" have directed the evolution of man and man's destiny through transhumanism, the Singularity, androids, artificial intelligence, and what H. G. Wells called the *World Brain* in his book of that name, or the "hive mind." This quantum leap in society occurred in America when the United States secretly imported thousands of Nazi rocket-, genetic-, and mind-control scientists as part of Operation Paperclip.

Aldous Huxley, Dr. Gregory Bateson, and others introduced the idea of the "scientific dictatorship" right out of Huxley's *Brave New World*, where a future totalitarian state rules the masses through a genetically bred "scientific dictatorship" that uses psychedelic drugs so the population doesn't know that their thoughts are programmed, that they even live in such a dictatorship, or that they've been programmed to love their slavery.

Global Political Awakening

Brzezinski correctly observed that due to the Internet, social media, and related technologies, a global intellectual and spiritual revolution is growing with such speed that a mass awakening is now occurring regarding what is happening. In 2009, Brzezinski published an article and gave a speech at the London-based Chatham House, or the Royal Institute of International Affairs, whose sister institute is the Council on Foreign Relations in the United States.

"For the first time in history almost all of humanity is politically activated, politically conscious and politically interactive," Brzezinski said. "It is, in essence, this massive 'global political awakening' which presents the gravest and greatest challenge to the organized powers of globalization and the global political economy: nation-states, multinational corporations and banks, central banks, international organizations, military, intelligence, media and academic institutions—the Transnational Capitalist Class (TCC), or 'Superclass'."

Brzezinski continued to talk about the globalists elite, or the "Super-class," and their plans for a "global political economy," but stated that it is threatened by people using the Internet and social media who are becoming aware of what is happening. Notice again that Brzezinski publicly disclosed that the globalists elite have already implemented their "scientific dictatorship" and they fully intend to rule the earth, making the masses their slaves and hoarding all the wealth and technology, including transhumanist technology.

Despite the endless rhetoric from the humanists and transhumanists elite about humanist ethics, compassion, caring, and science for the betterment of mankind, the documented writings and actions of many humanists and the most influential leaders reveal that they will continue to operate from the position of the Darwinian belief in "might makes right."

Below are the words of humanist Bertrand Russell as quoted from his 1985 book, *The Impact of Science on Society*. Notice that he openly states that, "Really high-minded people are indifferent to suffering, especially the suffering of others." What Russell means is that for himself and other members of the genetically superior elite, the common man is viewed like an ant.

> At present the population of the world is increasing…. War so far has had no great effect on this increase…. I do not pretend that birth control is the only way in which population can be kept from increasing. There are others…. If a Black Death could be spread throughout the world once in every generation, survivors could procreate freely without making the world too full… the state of affairs might be somewhat unpleasant, but what of it? Really high-minded people are indifferent to suffering, especially that of others.

Russell is not advocating the use of transhumanist technology to cure sickness, extend life, or create paradise on earth for the masses. He and other members of the elite plan to "cull the herd" and kill off all the "useless eaters" through man-made plagues developed in laboratories and spread through vaccines.

A vast array of medical sciences and technologies employs trans-humanist technologies and genetic engineering. But its purpose is to shorten life spans, cause people to die earlier, and scientifically dumb them down in the manner Aldous Huxley predicted in the *Brave New World*.

Rational Humanism

It should be noted that, beginning with the Age of Enlightenment and the Renaissance, and later fueled by the ideas of prominent humanists such as Julian Huxley, Aldous Huxley, H. G. Wells and Bertrand Russell, this global "scientific dictatorship" directly flows from what humanists call "rational humanism" and the elites' self-entitled claim of genetic superiority.

Prominent humanist leaders such as Julian Huxley, Aldous Huxley, H. G. Wells, Brzezinski, and others have openly written about their genetic and Darwinian right to use cruelty, brute force, torture, pain, and mass death to accomplish their ends. Russell like the Huxleys and others involved in the "scientific dictatorship," wrote openly about their plans. When Russell wrote, "Really high-minded people are indifferent to suffering, especially that of others," he was clearly saying that the elite are above things like traditional biblical morality, and as genetically superior beings, they have the evolutionary right to cast aside all notions of right or wrong. Many other members of the elite have said the same thing.

The body of literary evidence is overwhelming and proves that a significantly large percent of the elite, including authors, scientists, artists, politicians, international bankers, and others are calling for their right to enslave and kill mankind. Once again, note Brzezinski's words: "The technotronic era involves the gradual appearance of a more controlled society. Such a society would be dominated by an elite, unrestrained by traditional values."

Like many elites, Russell, Brzezinski, and others do not believe in the traditional Judeo-Christian moral values of right and wrong, and they do not intend to be restrained by them. They reject biblical values such as "love your neighbor as yourself" and choose humanist values derived from Darwinian evolution where "might makes right."

At its core, it is these Darwinian beliefs that were behind Hitler and the Third Reich. Hitler and his top men believed they were part of a master race of genetically superior beings, and as such, they had no problem killing off over eight million people in the concentration camps, including Jews, because they believed them to be genetically inferior. In addition, the Nazi scientists were using eugenics to genetically breed a master race, which is a transhumanist concept.

Irrational Humanism and Transhumanism

An obvious conflict exists between the compassion, human rights, human dignity, and human freedoms written in the Humanist Manifestos I, II and III, and the reality of all humanist-based revolutions beginning with the French Revolution.

One of the primary goals of humanism and transhumanism is to create a utopia or a transhumanist heaven on earth using science and technology. These same humanist promises echo throughout transhumanist writings such as Nick Bostrom's "The Transhumanist Declaration."

In this document, Bostrom calls for using science and technology to "end involuntary suffering," "the preservation of life and health," "the alleviation of grave human suffering," "policymaking ought to be guided by responsible and inclusive moral vision...respecting autonomy, individual rights…dignity of all people around the globe… moral responsibilities towards generations that will exist in the future." Bostrom then advocates the well-being of all life forms, including humans and favoring personal choice.

These goals of transhumanism are commendable, but there is a major disconnect here, because the past and present leaders of humanism and transhumanism are often the same very powerful people calling for a totally controlled totalitarian state where these sciences and technologies will be used to kill off, enslave, and dominate with unspeakable cruelty the lives of billions of people.

Simply reread the words of Julian Huxley, who first used the term "transhumanism." Read about the transhumanist dictatorship that Aldous Huxley called for in a *Brave New World*, which was not written as a warning, but as a future vision for the world. Also, recall Bertrand Russell's goals for the New World Order and Brzezinski's thundering voice calling for the use of science and technology to dominate and control the masses like subhuman slaves.

The reality is that, on a personal level, humanists and transhumanists are often the most ethical, loving, caring, and compassionate people I have encountered. My mother is a strong humanist—and many of her humanistic friends have more integrity and put love into action far more than many evangelical and fundamentalists I've met. The reality is that both humanists and evangelical Christians cannot be judged superficially or as a whole.

Millions of evangelical and fundamentalist Christians are also truly loving, compassionate, demonstrate integrity, and live to a higher moral code. Obviously, in many areas, humanists and Christians have strong differences as to what true morality is all about, but I'm talking about the areas in which Christians and humanists have common ground.

Bertrand Russell, an atheist and humanist, based his rejection of Christianity on the hypocrisy of Christians, but that argument goes both ways, because large numbers of humanists give lip service to things such as social justice, compassion for those that are hurting and the environment.

However, on a foundational level, humanism and transhumanism are built on the foundation of Darwin's theory of evolution, in which "might makes right" and only the fittest survive. This implies that things like power, intelligence, science, technology, and genetic superiority create an inevitable class system, or caste system, which Huxley wrote about in *Brave New World*.

In distinct contrast, Christianity is based on the reality of an infinite, personal, Living God of the Universe who sent mankind a Savior who is Jesus Christ.

The biblical God is a God of love, and Jesus Christ died on a cross in real space, time, and history to remove the death force, or sin, from anyone who chooses to believe in Jesus Christ and have faith in what He did.

Humanists dismiss the Bible and the account of Christ as being nothing more than a fairy tale with no scientific or empirical evidence to support it. But there is a great deal of documented historical evidence and scientific empirical evidence to support such claims, including archeological evidence, numerous fulfilled prophecies in history, and multiple eyewitness accounts of the resurrection of Jesus Christ that would hold up in any courtroom in England.

A Lack of Scientific Evidence to Support Humanism's Claims

Despite its claims to be scientific, rational, and based on empirical evidence, humanism and transhumanism are based on faith just as much as Christianity is, and thus constitute a "new religion," as Julian Huxley and its founders claimed.

The religious doctrine of humanism teaches that man randomly evolved from nothing—just from pure chaos and random chance. As such, life has no meaning, because man is the product of Darwin's theory of evolution despite that there is no documented, empirical evidence to prove that evolution is true. Ultimately, it's believed through irrational and religious zealotry and an "upper-story leap" into a mysticism that arbitrarily rejects reason and science.

Darwin's theory of evolution, taken literally and at its plain meaning communicated by its intellectual, scientific, and philosophical supporters, states that life has absolutely no meaning, except what a person chooses to create, and there is no God or absolute right or wrong. Evolutionary theory clearly states that the entire process of evolution, which would include transhumanism, is based on a biological mechanism in which "only the fittest survive" and brute force and power, or "might makes right," are the "final reality."

Those who see this clearly, and not through rose-colored glasses, know that this means that because there is no meaning to life, or a personal God, then ultimately every man or woman has the evolutionary right to do absolutely anything, if he or she has the power or brute force to get away with it.

Since transhumanism and humanism don't recognize that man has a fallen nature, they have no proper checks and balances to restrain evil, and that's why the elite totalitarians are in the process of dominating the world with cruelty and force that are unimaginable to transhumanists and humanists who possess a self-imposed moral code.

Just like the irrational idealism of the counterculture of the 1970s, which promised love and peace, but ended at a rock 'n' roll concert at the Altamonte Speedway in California—where the Rolling Stones were performing "Sympathy for the Devil" onstage and an outlaw motorcycle gang, Hells Angels, murdered a man near the stage as documentary filmmakers captured his killing on multiple cameras. Peace and love faded quickly with this murder, and the peaceful hippies were replaced by the violent revolutionaries known as the Weather Underground.

The Achilles Heel of Humanism
and Transhumanism

The basic foundational flaw of both humanism and transhumanism is that they refuse to recognize the reality that men and women were created in the image of the infinite, personal, Living God of the Universe and were given the DNA of God.

The very things Sir Francis Bacon and transhumanists seek, which is to empower mankind to self-evolve through science and technology so they can rule over and have dominion over life, is what God gave Adam and Eve in the Garden of Eden, where they were given the power to rule and reign over their lives and the earth, along with being given dominion.

These powers, along with the fact that they did not age or get sick and were immortal beings, were lost when Adam and Eve rejected the one commandment from the Creator to not eat from the fruit of the tree in the middle of the garden. They were tempted by an interdimensional being the Bible calls Lucifer, who came to them in the form of an upright, reptilian being, and the temptation centered around his promise that, "you shall be as gods."

However, as soon as Adam and Eve ate of the fruit, the death force entered their beings and all of creation. Instantly, they knew they were dying and had lost their supernatural ability to rule and reign. In addition, this death force, which the Bible calls the law of sin and death, caused them to manifest a fallen human nature, which at its core was dark, evil and in rebellion against God.

It is not that men and women lost their ability to be loving, kind, and compassionate, and to accomplish remarkable achievements such as science and technology, it is simply that the death force operating in them continually releases, as the movie *Stars Wars* contends, "the dark side of the force."

The definitive proof regarding the Fall of Man is death itself. Both Christians and transhumanists recognize that death is something to be conquered; it's just that they have radically different approaches as to how to accomplish this.

The Death Force

At one time in my life, I believed the whole notion of sin was ridiculous—an archaic concept left over from the religious Middle Ages.

I also didn't believe in evil, per se, or demons, or a personal devil. I thought that was laughable, pathetic, and a means for religious authoritarianism to repress men and women's creativity, free thought, behavior, sexuality, and ability to use science and reason to create a utopian society.

However, based on observable, measurable, and historical evidence, I believe all men and women have deep within them a DNA-directed propensity toward thoughts, actions, and behaviors that create death and oppression on every level.

When mankind was originally created with the DNA of God, they existed in their optimal and enhanced state. But when the Fall of Man happened and the death force entered the first man and woman, their DNA was reconfigured by the death force. As a result, every man or woman born since that time contains the DNA of God, but in a highly degraded form that is passed on generationally.

What scientific empirical evidence of the existence of a death force do we have? We do not have sufficient scientific evidence to quantify this death force. On a personal level, though, I believe one day soon we will have the technology to reveal the reality of this death force.

Throughout man's history, many things were real, but we didn't have the technology to prove them. For example, scalar technology is real, and we now know this because we have the technology to use and measure it. The same would be true of the ten different dimensions proven by quantum physics and string theory, electromagnetic waves, bioenergetic fields, biophotons, and other discoveries.

The Engineered Counterculture, Plato's Atlantis, and Unspeakable Horrors

The Fall of Man is the explanation as to why transhumanism's attempt to create artificial immortality and a scientifically engineered heaven on earth cannot work in the long term. As I related my experiences with the counterculture, it was not until decades later after enormous research that I realized that the entire counterculture movement, along with "peace, love, sex, drugs, and rock 'n' roll," were not some spontaneous cultural movement.

This movement was designed and created by the hidden hands of Aldous Huxley's "scientific dictatorship" and people like Bertrand Russell, Dr. Gregory Bateson, Dr. Timothy Leary, Dr. John C. Lilly, the Stanford Research Institute, the Tavistock Institute, Ken Kesey, and others who were backed and financed by British and American intelligence.

Aldous Huxley came to the United States in 1936 to spread mescaline among the Hollywood entertainment industry and returned years later to influence Project MKULTRA and to consolidate the "scientific dictatorship" in America, which he openly spoke about in the early 1970s at a University of California, Berkeley gathering of prominent neuropsychiatrists.

The reality is that there are two Aldous Huxleys. One is known as a compassionate intellectual who wanted to use transhumanism, science, and technology to help all of mankind. The second is the Huxley who openly wrote and spoke about enslaving mankind.

This dichotomy is apparent among numerous humanist and transhumanist leaders. The reason for this is that elite humanists and transhumanists believe strongly in part of Plato's *Republic,* in which he wrote that the legendary Atlantis was real and was ruled by ten philosopher-kings, or a scientific elite, for the good of the masses.

It's interesting that both the Enlightenment and the French Revolution provided an idealized utopian vision of mankind's future by ignoring reality and adopting a humanistic or transhumanist mysticism that was in direct conflict with reason.

The lofty humanistic ideas of the French Revolution, which paraded the Goddess of Reason throughout the streets of Paris, the belief in the enlightened or illuminated man, ended in a bloodbath and mass beheadings in the streets of Paris. The "scientific humanism" of the French Revolution birthed the socialist, Marxist, and communist revolutions that not only did not produce a utopian state, a "workers' paradise," a fair redistribution of wealth, and unprecedented human freedom, but every Marxist and communist revolution, without exception in history, has produced a series of bloody revolutions in communist or Marxist nations that ended up killing over two hundred million people and sentenced millions to re-education camps, mental hospitals, and unspeakable torture.

The "Upper-Story Leap" of Totalitarian States

These scientific-Marxist societies not only didn't produce economic prosperity or the fair redistribution of wealth, but they stripped people of their freedoms, lowered their standards of living, and created oppressive totalitarian states.

However, when we read the Humanist Manifestos I, II, and III, along with the transhumanist declarations, we consistently see a denial of the obvious historical empirical evidence about such totalitarian states, but we see this consistent "upper story leap" into a kind of mystical-humanistic-optimism that is insanely irrational about the nature of all totalitarian states.

Despite the self-criticism within the Humanist Manifestos by Julian Huxley and others admitting to the dangers of communism and Marxism, there still exists an irrational romanticism about a utopian and communist-socialist world state.

This romanticism denies the empirical evidence of history, rejects reason, and takes us into an "upper-story leap" into mysticism. This is often evidenced by a mystical and romantic longing for the United Nations as the possible ideal humanistic state or world socialist state that will usher in paradise on earth under the new religion of man, which the book of Revelation says will be ruled by the Antichrist.

The late Beatle John Lennon's definitive song, *Imagine*, which is sung around the world on New Year's Eve as the Times Square ball drops, is an anthem to "when the world will be as one," as he tragically looked through rose-colored glasses and "imagined" there would be a world without money, borders, heaven and hell, religion, and what *Ode to Joy* poet Friedrich von Schiller called the "brotherhood of man."

Tragically, Lennon was shot to death outside of his Dakota building apartment—the same building where director Roman Polanski shot the movie *Rosemary's Baby* with Mia Farrow and actress Sharon Tate.

Rosemary's Baby was an occult, transhumanist baby born of the DNA of a human and demon in Aleister Crowley's "Moonchild Ritual." Polanski and his wife, Sharon Tate, were alleged to have been part of some occult group, and she was brutally stabbed to death in the Hollywood Hills area by the Manson family followers of Charles Manson. Lennon's tragic death and Tate's death killed the romanticism.

Humanism and Transhumanism as the New Religion

Ultimately, humanism and transhumanism are simply the new religion because many of their "doctrines of theology" require us to believe or have faith apart from reason and empirical scientific evidence to believe things such as Darwin's theory of evolution.

A further proof that humanism is irrational is the fact that, in the end, humanism wants to define itself as a new kind of religion, which is precisely the opposite of what humanism originally defined itself as, which was not a religion, but a product of science and reason. In "The Coming of the New Religion of Humanism" in *The Humanist* in 1962, Julian Huxley wrote:

> Much of every religion is aimed at the discovery and safeguarding of divinity, and seeks contact and communion with what is regarded as divine. A humanist-based religion must redefine divinity, strip the divine of the theistic qualities which man has anthropomorphically projected into it, search for its habitations in every aspect of existence, elicit it, and establish fruitful contact with its manifestations. Divinity is the chief raw material out of which gods have been fashioned. Today we must melt down the gods and refashion the material into new and effective agencies, enabling man to exist freely and fully on the spiritual level as well as on the material.

Julian Huxley's statement is based on an "upper-story leap" from reason to mysticism, because he is simply replacing the Christian God with man as God. This was the temptation given by Lucifer to Adam and Eve: "You shall be as gods." Despite what Huxley says, people have fashioned themselves into gods. Huxley continues:

> The character of all religions depends primarily on the pattern of its supporting framework of ideas, its theology in an extended sense.... I feel sure that the world will see the birth of a new religion based on what I have called evolutionary humanism. Just how it will develop and flower no one knows—but some of its underlying beliefs are beginning to emerge, and in any case, it is clear that a humanism of this sort can provide powerful religious, moral and practical motivation for life.

Julian Huxley, the Apostle Paul of humanism, sought to provide for man what the Christian religion does—which is irrational, because humanism is supposed to be based on scientific evidence and not faith.

Finally, it must be understood that transhumanism is absolutely a religion. Both humanism and transhumanism, which are largely based on faith and not reason or science, cleverly employ rhetorical language to say, something to the effect of, "transhumanism is the new religion and it is the only religion which is true because it is based on scientific fact, empirical evidence and reason." This is nothing more than the apologetics of humanism, just as Christianity, Judaism, and Islam maintain apologetics. It is simply saying our religion, transhumanism, is the only true religion, and therefore you should reject all others. What we are witnessing is humanism and transhumanism preaching the new gospel of transhumanism through evangelism.

But, in a truly credible theological debate, both sides must have documented evidence that their religion is true. The religion of transhumanism has zero documented evidence to even remotely prove that Darwin's theory of evolution is true, and that is the foundational tenet of the religion of humanism and transhumanism.

Knowing that they have zero empirical evidence regarding evolution, they then require us to have faith or belief in humanism and transhumanism, which means that their religion is not based on reason, science, and empirical evidence, as they vigorously maintain. A lot of intellectual and rhetorical tap dancing is being done here.

In distinct contrast, the religion of Christianity has countless scientific and documented historical facts to support its claims, such as the numerous detailed prophecies that have come true and a historical document, the Bible, which accurately provides dates and witnesses regarding the many prophecies that have already come true.

Two thousand years ago, the Jews were driven from Israel after their temple was destroyed, and they were dispersed throughout the world. Jesus Christ predicted this to His disciples. Numerous Old Testament prophets, including Daniel, Ezekiel, Joel, Jeremiah, and others predicted that the Jews would return to the physical land of Israel in the last days, which happened in 1948.

Additionally, there are numerous eyewitness accounts of the death and resurrection of Jesus Christ, along with detailed prophecies regarding His virgin birth and where He would be born. The tomb of Christ was surrounded by the Roman equivalent of "special forces," and so there was no possibility His body was stolen. Finally, again there were numerous eyewitness accounts of people seeing and talking to Jesus in His new glorified body. These are just a tiny number of proofs for the claims of Christianity.

It's impossible to avoid the fact that, at its root, transhumanism is a religion that offers synthetic or virtual salvation in the form of uploading the human consciousness into a cloned body, humanoid, android, cyborg or robot, like the movie *Transcendence* starring Johnny Depp.

Transhumanism offers to build heaven on earth through a scientific, global utopia, but it ignores the empirical evidence that every single humanist attempt at building paradise on earth has resulted in a nightmare.

The global government of ancient Babylon at the Tower of Babel failed because the personal, living God of the Universe saw the evil that would be produced through this "Mystery, Babylon" system. Other humanist efforts at creating a utopia or paradise have horribly failed, from the French Revolution to the Bolshevik or Russian communist revolution to every other communist revolution.

The final humanistic revolution all around us is the globalists elite's plan for a one-world government, but it is embedded with the same totalitarian neurological virus as both the French Revolution and communism.

The disparity between what humanism and transhumanism claim to be versus what they produce can only be understood when we acknowledge that humanism is irrational, and it is not based on reason or scientific empirical evidence, but requires faith to believe, just like every religion. Transhumanism is a religion that simply changes gods. Instead of the God of the Bible, transhumanists worship man as god.

The Fight That Shows No Pity
Christian Transhumanists and the Quest of Gilgamesh

By Milieu Member Derek Gilbert

For mankind, whatever life it has, be not sick at heart,
be not in despair, be not heart-stricken!
The bane of mankind is thus come, I have told you,
what was fixed when your navel-cord was cut is thus come,
    I have told you.
The darkest day of mortal man has caught up with you,
the solitary place of mortal man has caught up with you,
the flood-wave that cannot be breasted has caught up with you,
the battle that cannot be fled has caught up with you,
the combat that cannot be matched has caught up with you,
the fight that shows no pity has caught up with you!
—*Epic of Gilgamesh*

More than five thousand years ago, the legendary Sumerian King Gilgamesh embarked on a single-minded quest to procure the secret of immortality. According to the story, he was so distressed by the death of his best friend, Enkidu, and obsessed with overcoming his own mortality, that he tracked down the Sumerian Noah, Utnapishtim the Far-away, for advice.

A thousand years or so after Gilgamesh succumbed to the fate that awaits us all, around 2000 BC, Amorites overwhelmed the native Sumerian and Akkadian rulers of the Fertile Crescent. By 1900 BC, the time of Abraham, Amorite dynasties controlled nearly every kingdom and city-state from the Persian Gulf to the Levant—modern Iraq, Syria, Jordan, Lebanon, and Israel. By the time of Jacob, Amorites had moved south and taken over northern Egypt, too.

To the pre-Flood magical and religious practices of the Sumerians, the Amorites added ancestor worship, but with a twist. From the evidence, some of which has only been found within the last hundred years and translated within the last forty, it appears that the kings of the Amorites believed they descended from the gods who ruled before the Flood—the ones who, in Babylonian, Hittite, and Greek cosmology (as well as the Bible),[187] were locked away in an underworld prison reserved for supernatural threats to the divine order.

Further, the Amorites, at least during the second millennium BC, performed rituals to summon the spirits of those gods to bless their kings.

Now, flash forward four thousand years to today: In the West, we've been so indoctrinated by positivism (a philosophy that teaches science is the only reliable tool for finding truth) that we're more likely to believe the old gods were alien astronauts than supernatural beings. But the quest to unlock the secret of immortality continues. The modern transhumanist movement holds out the same promise offered to Adam and Eve in Eden: To paraphrase, "Ye shall not surely die; your eyes shall be opened, and ye shall be as gods."

As Christians, we reject positivism. That's not to say we're anti-science, but admittance to our club requires believing (among other things) that an invisible, all-powerful deity spoke everything into existence; that He manifested as a fully human man in a dusty, backwater province of the Roman empire about two thousand years ago; that He died for our sins; and then that He, three days later, literally rose from the dead. There is no way to syncretize Christian faith with a philosophy that rejects theology and the metaphysical. And yet that's exactly what transhumanists are trying to do—even if most of them don't realize it or won't admit it.

Transhumanism is a growing movement that wants to fundamentally transform human physiology through cutting-edge technology, with the goal of achieving eternal life through science. In other words, transhumanists are trying to weld together two diametrically opposed worldviews—one based on the supernatural and the other that denies its existence.

Of all people, then, it's surprising to find that some Christians are making common cause with transhumanists. By so doing, these Christians are unwittingly summoning those old gods and offering them one last shot at knocking God off His mount of assembly.

That's why The Milieu is speaking up.

Two generations before Gilgamesh, a Sumerian king named Enmerkar ruled the ancient Near East. Both men ruled from the city of Uruk in what is today southeastern Iraq—which, you might have noticed, is just a different spelling of the city's name. In the Bible, it's spelled a third way—Erech, which, along with Babel, was "the beginning of [Nimrod's] kingdom." From there, the legendary kings of Uruk ruled nearly the entire Fertile Crescent, the land between the Euphrates and Tigris rivers in what is now Iraq, Syria, and southern Turkey. Scholars call this the Uruk Expansion, a period between about 4000 BC and 3100 BC. Logically, Nimrod and Gilgamesh would fit somewhere in that time frame.

Like Nimrod, Enmerkar was the second generation after the Flood. In my book *The Great Inception*, I show why Babel is not to be confused with Babylon, which wasn't founded until at least a thousand years after the tower's construction was interrupted, and make the case that Enmerkar and Nimrod were one and the same. Babel was most probably at the ancient city of Eridu, and the tower was a ziggurat, the largest and oldest ever found in Mesopotamia, built as a temple to the Sumerian god Enki, the lord of the abyss.

According to a poem from the time of Abraham, Enmerkar/Nimrod hoped to build up the temple of Enki into a "holy mountain," and to "make the great abode, the abode of the gods, famous for me."[188] An abode of the gods directly above the *abzu*, the abyss? No wonder YHWH decided to personally intervene!

The Sumerian King List names Lugalbanda as Enmerkar's successor as king of Uruk, and he was succeeded in turn by Gilgamesh. We don't know whether Gilgamesh was Enmerkar/Nimrod's grandson, but scholars generally consider him a historical character. A team of German archaeologists mapped Uruk in 2001 and 2002 using cesium magnetometry, and among their discoveries was a building under what was the bed of the Euphrates River in the third millennium BC that *might* be the burial crypt of the legendary king.[189]

We don't know whether Gilgamesh was Nimrod's grandson, but he had his predecessor's ambition and then some. Where Nimrod tried to conquer the known world and build a home for the gods in his kingdom, Gilgamesh set his sights on becoming immortal.

Evidence suggests that the king may have resorted to bringing back knowledge that had been lost beneath the waters of the Great Flood. According to the Book of Enoch, a group of angelic beings, called Watchers by the Hebrews, descended to the summit of Mount Hermon in the days of the patriarch Jared.[190] As Dr. Michael Heiser noted in *Reversing Hermon*, there was more to the visit of the Watchers than producing monstrous offspring; the rebellious angels brought with them information mankind was not meant to possess: Sorcery, charms, the cutting of roots and plants (probably for mixing potions), metalworking and the making of weapons, makeup (and presumably the art of seduction), and reading fortunes in the movement of the stars. In short, the Watchers lured humanity into evil, and "all the earth was filled with the godlessness and violence that had befallen it."[191]

Gilgamesh was referred to on a Mesopotamian cylinder seal as "master of the *apkallu*,"[192] and by the time of Hammurabi the Great, who was probably a contemporary of Isaac and Jacob, Gilgamesh was viewed as the one who had returned to mankind the pre-Flood knowledge of the *apkallus*—the Mesopotamian name for the Watchers.[193] In fact, it appears the sages and priests of Babylon believed it was precisely that arcane knowledge which (to borrow a phrase) Made Babylon Great Again.

Interestingly, the Old Babylonian text of the Gilgamesh epic establishes another link between Gilgamesh and the Watchers. To make a name for himself, Gilgamesh and his drinking buddy Enkidu decided to kill Huwawa (or Humbaba), the monster who guarded the Cedar Forest. In a sense, the pair aimed for a sort of immortality by performing a great deed.

> Hear me, O elders of Uruk-the-Town-Square!
> I would tread the path to ferocious Huwawa,
> I would see the god, of whom men talk,
> whose name the lands do constantly repeat.
> I will conquer him in the Forest of Cedar:
> let the land learn Uruk's offshoot is mighty!
> Let me start out, I will cut down the cedar,
> I will establish forever a name eternal![194]

The Old Babylonian text of the epic locates the cedar forest on the peaks of "Hermon and Lebanon."[195] After killing Huwawa, the two friends "penetrated into the forest, opened the secret dwelling of the Anunnaki."[196]

This is significant for a couple reasons. First, the mission of Gilgamesh and Enkidu may have been far darker than it appears on the surface. The late Dr. David Livingston, founder of Associates for Biblical Research, pointed out that "Huwawa" may have sounded a lot like "Yahweh" in ancient tongues. If Livingston was right, then the *real* mission of Gilgamesh was to achieve immortal fame and glory by killing the guardian of the secret home of the gods—Yahweh.[197]

Secondly, the Anunnaki, who were originally the great gods of Mesopotamia, had become the gods of the underworld by the time of Abraham.[198] Marduk, after defeating the chaos dragon Tiamat, decreed that the Anunnaki, or at least half of them, should relocate permanently to the nether realm.[199] The Hittites, who lived north of Mesopotamia in what is now Turkey, identified the Anunnaki as primordial deities of the underworld, possibly "an earlier generation of gods who had retired or were banished by the younger gods now in charge."[200]

This is relevant because Gilgamesh, despite his desperate effort to avoid "the bane of mankind," died anyway—and upon his death, according to the legend, was made ruler of the dead.

Gilgamesh, in the form of his ghost, dead in the underworld, shall be the governor of the Netherworld, chief of the shades![201]

This has special significance because of the importance of the ancestor cult among the Amorites, who founded the old kingdom of Babylon. For more than a thousand years, Amorites in the ancient Near East (modern Iraq, Syria, Lebanon, Jordan, Israel, Saudi Arabia, and northern Egypt) venerated their dead, especially the dead ancestors of their kings.[202] Although Gilgamesh was a Sumerian king who had departed this world a millennium before the great kings of Babylon, it seems he epitomized the venerated royal dead, and he played an important role in the ancestor cult and magical healing rituals of Babylon.[203]

In *The Great Inception*, I quote Canaanite (western Amorite) texts that describe rituals to summon the Rephaim—the spirits of the Nephilim (note that the Hebrew word *rapha* is sometimes rendered "shades" in the Bible)—and something called the Council of the Didanu, which was apparently an underworld assembly of the old gods.

Now, get this: *Didanu* was the name of an ancient Amorite tribe from which the kings of Babylon, old Assyria, and Canaan claimed descent, and—here's the good part—it was the word from which the Greeks got the name of their former gods, the Titans.[204]

Pause for that to sink in. Kings of the Amorites, neighbors of the ancient Hebrews from the time of Abraham through the time of the Judges, apparently believed they descended from gods later known to the Greeks as the Titans—the elder generation of deities who were overthrown by Zeus and the Olympians and banished to Tartarus.

Let's take a moment to stop and summarize here. This is starting to make my head spin, and I'm the one writing.

- Gilgamesh, a legendary (but probably historical) post-Flood king of Uruk in the fourth millennium B.C., was obsessed with finding the key to immortality.
- He died anyway sometime around 3000 BC, give or take a few centuries.
- Amorites more than a thousand years later linked Gilgamesh with the "shades" (the Rephaim?), the *apkallu* (the Watchers/ Titans), and the Anunnaki, the gods of the underworld.
- If the Hittites were correct in identifying the Anunnaki as "former gods" who'd been overthrown and banished to the netherworld, then they, too, can be identified as the Hebrew Watchers and Greek Titans.
- The Anunnaki and the Watchers (and thus the Titans) were linked to Mount Hermon. Mount Hermon is also where Gilgamesh and Enkidu killed the monstrous Huwawa.
- By comparing their stories with the Bible, we can identify the Titans, the Anunnaki, and the *apkallu* as the Watchers of Genesis 6, "the angels who did not stay within their own position of authority" who are "kept in eternal chains under gloomy darkness until the judgment of the great day."[205]

All well and good, you might say, but what does any of that ancient history have to do with modern transhumanism, and especially with Christians who call themselves transhumanists? Glad you asked.

Without realizing it, today's transhumanists are replicating the quest of Gilgamesh for the secret of immortality. Most transhumanists are atheists, which makes their mission easier to understand. (Although Gilgamesh wasn't an atheist, the Mesopotamian afterlife couldn't have been much fun if he was driven to such lengths to find a way to avoid it.)

Christian transhumanists, on the other hand, are more like Adam and Eve, or those who lived during the time of the Watchers' descent—people who should have known better (Adam lived another 470 years after the birth of Jared), but who willingly traded away their lives for secrets that would make them like gods. Or so they thought.

First things first: Your view of end-times prophecy has a powerful effect on how you live out your Christian faith. For example, if you believe that Jesus won't return until the end of the millennial reign prophesied in Revelation 20, or that the thousand-year reign is symbolic rather than literal, then it's understandable that you might believe a Christian's duty is to work toward the creation of heaven on earth.

Amillennialism, which teaches the latter view, is the majority view among the world's Christians. It is the official position of the largest Christian denominations—Roman Catholicism, the Eastern and Oriental Orthodox churches, and some of the mainline Protestant denominations. While many of the believers in the Catholic and Orthodox churches hold a supernatural worldview, it's no surprise that the combination of a scientistic culture and an eschatology that foresees the world getting better and better until Jesus returns would produce a subset of Christians who accept, at least in part, the philosophy of the transhumanists.

So, what is transhumanism, exactly? According to leading bioethicist Wesley J. Smith, it is "an emerging social movement that promotes the technological enhancement of human capacities toward the end of creating a utopian era in which 'post humans' will enjoy absolute morphological freedom and live for thousands of years."[206]

Transhumanists themselves are more direct in describing their goals and the means of attaining them:

> Biology mandates not only very limited durability, death and poor memory retention, but also limited speed of communication, transportation, learning, interaction and evolution... transhumanists everywhere must support the revolutionary movement against death and the existing biological order of things.[207]

In short, transhumanists believe God's design is inherently flawed, so we humans must get busy "speeding up evolution and becoming true masters of our destiny."[208]

Zoltan Istvan, who ran for president in 2016 as a candidate for the Transhumanist Party (which he founded), distilled the objectives of transhumanism into a philosophy he calls Teleological Egocentric Functionalism. Istvan summarized those ideas with his Three Laws of Transhumanism:

1. A transhumanist must safeguard one's own existence above all else.
2. A transhumanist must strive to achieve omnipotence as expediently as possible—so long as one's actions do not conflict with the First Law.
3. A transhumanist must safeguard value in the universe—so long as one's actions do not conflict with the First and Second Laws.[209]

You probably noticed that the first of Istvan's laws runs head first into Jesus' description of true love: "Greater love hath no man than this, that a man lay down his life for his friends."[210]

It shouldn't surprise you that Istvan is an atheist. His three laws make perfect sense to those who believe there's no God and no existence whatsoever after death. Even the actions of Jethro Knights, the protagonist of Istvan's novel *The Transhumanist Wager*, are understandable, if chilling. To rid the world of the corrosive superstition of religion (as Istvan sees it), Knights and his team of super geniuses form a new, independent nation called Transhumania, and then they proceed to destroy the Vatican, the Kaaba in Mecca, and the Wailing Wall in Jerusalem, among other iconic religious and cultural sites around the world.

And Knights has no problem with adopting a eugenics program that would have made Hitler proud:

> We need to divert the resources to the genuinely gifted and qualified. To the achievers of society—the ones who pay your bills by their innovation, genius, and hard work. They will find the best way to the future. Not the losers of the world, or the mediocre, or the downtrodden, or the fearful. They will only drag us down, like they already have.[211]

To be fair, even other transhumanists are put off by *The Transhumanist Wager.*

> The Transhumanist Wager has only one idea—a fascistic interpretation of the meaning of transhumanism in which the complexity of every other current of human thinking, including transhumanism itself, is reduced to a cartoon.[212]

I bring this up because it's important to understand that transhumanism means different things to different people, and not all transhumanists are ready to exterminate the "unfit."

Obviously, whether one believes in God is a key factor in whether one thinks living forever through technology is a good idea. But theists

(like Christians) can have very different views of transhumanism ranging from acceptance to, well, The Milieu.

For example, there is a very active Mormon Transhumanist Association. This isn't entirely unexpected, since Mormon theology is about
apotheosis (at least for men)—literally becoming gods.[213] But while it's a
shorter leap to transhumanism from the doctrines of the Church of Latter-day Saints than from the New Testament, there are, as noted above,
a small, but growing, Christian Transhumanist Association.

It is difficult to see how any Christian who holds a futurist view
of prophecy can embrace transhumanism. That said, not all Christian
transhumanists are working toward virtual immortality and "absolute
morphological freedom"—changing one's body shape, physical abilities,
and gender at will. Many of them appear to be concerned with just
improving the quality of life for the poor and downtrodden, seeing this
as a mission that follows in the footsteps of Jesus.

And yet well-meaning Christians who ally themselves with the transhumanist movement, because they believe advances in genetics, robotics, artificial intelligence, and nanotechnology will yield better tools for
fulfilling this mission, are giving tacit approval to an alternate religion
that promises salvation through technology. By promising to make us
more than human, transhumanists and their Christian allies declare that
God's design is imperfect and must be improved.

By us. Sure. Because we're smarter and have better technology than God.

In other words, in their effort to make humanity into H+, they will,
ironically, downgrade the species to H-. And there is no guarantee that
if they succeed to any degree, Humanity 2.0 will still be "human" in any
meaningful way.

Christian transhumanists have good intentions. The Christian Transhumanist Association describes its mission as follows:

As Christian Transhumanists, we seek to use science & technology to participate in God's redemptive purposes, to cultivate life and renew creation.

1. We believe that God's mission involves the transformation and renewal of creation including humanity, and that we are called by Christ to participate in that mission: working against illness, hunger, oppression, injustice, and death.
2. We seek growth and progress along every dimension of our humanity: spiritual, physical, emotional, mental— and at all levels: individual, community, society, world.
3. We recognize science and technology as tangible expressions of our God-given impulse to explore and discover and as a natural outgrowth of being created in the image of God.
4. We are guided by Jesus' greatest commands to "Love the Lord your God with all your heart, soul, mind, and strength…and love your neighbor as yourself."
5. We believe that the intentional use of technology, coupled with following Christ, will empower us to become more human across the scope of what it means to be creatures in the image of God.[214]

Analyzing the CTA's affirmation runs the risk of setting up a series of straw men, and we don't have any intention of doing that if we can help it. How we respond to their five points depends on how we define our terms.

The opening statement sounds harmless enough, but what exactly is meant by participating in "God's redemptive purposes"? If we accept the definition of the word "redemptive" as "acting to save someone from error or evil," then transhumanism, properly directed, is a means toward that end. Who can look on a child suffering from, say, a debilitating injury or genetic defect of any type without thinking it an error or evil? Wouldn't any of us use whatever was in our power to correct such a wrong?

But how do we define "error" or "evil"? If we interpret God's redemptive purpose according to the straightforward Gospel that Paul gave to the church at Corinth "as of first importance"—that "Christ died for our sins in accordance with the Scriptures, that he was buried, that he was raised on the third day in accordance with the Scriptures"[215]—then the evil from which we are redeemed is spiritual, even if the physical, emotional, and mental imperfections that plague our world are sometimes side effects of that evil. In other words, what the Christian Transhumanist Association calls "the transformation and renewal of creation" needs sharper definition before we can sign on to that mission.

And we are certainly sympathetic to "working against illness, hunger, oppression, injustice, and death." Jesus healed the sick and fed the hungry, and the Bible is clear about a Christian's duty to help the poor, orphans, and widows. But Jesus also came "to destroy the works of the devil."[216] He cast out a lot of demons, and his choice of Mount Hermon for the Transfiguration was deliberate.[217] We are in the middle of a very long spiritual war, and God isn't called Lord of Hosts—i.e., Lord of *Armies*—for nothing.

Jesus didn't say, "The poor you always have with you," because there aren't enough charities, but because evil is the default setting of the human heart. There will always be men and women who enrich themselves at the expense of others.

That is the evil from which the world must be redeemed. And Christ is the Redeemer—not you and me.

That's the heart of the matter. What are Christian transhumanists trying to accomplish? While they embrace science, technology, and the full engagement of our minds in the service of our Lord, they may fail to recognize the danger in going too far. Scientists and tech entrepreneurs like Stephen Hawking and Elon Musk have likewise sounded warnings about the dangers of runaway science; in fact, Musk described the development of artificial intelligence, "our biggest existential threat," as "summoning the demon."[218]

The Rev. Dr. Christopher J. Benek, founding chair of the Christian Transhumanist Association, acknowledges that "the practical outcomes of escapist concepts in science and technology are just as bad as they are in theology."[219] But, like Hawking and Musk, he argues that we must push forward anyway:

> Christians are called to help in Christ's redemptive purposes on earth as we seek to actualize Christ's great prayer "on Earth as it is in Heaven."
>
> The reality of humanity is that humans are called to be CoCreators with God and one another. We have a divine appointment to explore possibilities and discern what constitutes proper choices in order to appropriately steward science and technology.[220]

Well, we can agree in part. Yes, we Christians have a duty to appropriately steward science and technology. That's why I'm proud to be called a part of The Milieu, and why I'm writing this instead of watching *Doctor Who*.

But "CoCreators with God"?

Whoa, there. That's where we run into trouble. That's where hubris can lead to really spectacular mistakes. And while I don't think for a minute that Christian transhumanists are likely to do anything close to what secular transhumanists propose (like merging human biology with machinery to acquire godlike abilities), just describing the mission as co-creation with God moves humanity up a step in the cosmic order without His express written consent.

Frankly, the concept of co-creation moves uncomfortably close to territory occupied by Dominionists, a group within charismatic Christianity that believes we Christians must literally take over the world before Jesus can return. They, too, would describe their mission as participating in "Christ's redemptive purposes on earth." They believe we're called to defeat God's enemies and make them a footstool beneath Christ's feet. What could be more redemptive than that?

Christian transhumanists don't share that view. My concern is that their emphasis is on the wrong end of the rope. As noble as their motives are, they're pulling in the same direction as the rest of the transhumanists; which is to say, *Luciferian* transhumanists—an accurate description, whether they recognize it or not.

I don't mean that non-Christian transhumanists worship Satan. Most of them wouldn't even acknowledge the existence of Satan.[221] By "Luciferian," I mean they're committed to the goal of creating heaven on earth—what scholars would call "anthropocentric soteriology," or salvation by our own works.

Being Christians, I am confident that the Rev. Dr. Benek and his colleagues deny that their hope for salvation is in anything other than the atoning death of Jesus Christ. But then that begs the question: Why work for the "transformation and renewal of creation"? Dr. Benek rationalizes it thus:

> When Christ sent people out—to effectively do what seems miraculous from [a] given perspective in time—he was calling them to imagine a better way of looking at the world and then challenging them to believe that better way into existence with action. Now that technology is expanding the possibilities of our imaginations, we are faced with new opportunities to virtuously live into the wonder of these possibilities.[222]

But those weren't our marching orders. In the beginning, we were told to "be fruitful and multiply and fill the earth and subdue it."[223] Later, Jesus commanded His disciples to "go therefore and make disciples of all nations, baptizing them in the name of the Father and of the Son and of the Holy Spirit."[224] There was nothing in there about transforming and renewing. That's what the military calls "mission creep."

And it can be dangerous. First, it puts our focus on this world instead of the next. Christians have an absolute duty to demonstrate sacrificial love to those around us—but we should never, ever forget that the whole point of doing so is to open doors so that we can share the Gospel of Jesus Christ.

Second, we're on the verge of being able to permanently change what it means to be human at the genetic level. We are nearing a day when we can create an autonomous intelligence capable of thinking millions, if not billions, of times faster than us. It is misguided to join forces with those who not only want to put that kind of power into human hands, they want to redesign and fundamentally transform God's ultimate creation, humankind.

That is ironic. As Christians, we are all transhumanists by definition. To paraphrase what Paul wrote to the Corinthians, we won't all die but we will all be changed.[225] At His return, Jesus will grant all who have accepted Him as Lord the immortality we were originally created to enjoy.

Now, it's understandable that non-Christian transhumanists would reject that in favor of a Manhattan Project-scale effort to unlock the key to eternal life. By why on God's green earth would a Christian trade God's promise for an offer of artificial immortality?

A few final thoughts and some speculation from a prophetic perspective. If advances in genetics, robotics, artificial intelligence, and nanotechnology ever succeed in bringing virtually unlimited lifespans to humanity, it's not going to create the heaven on earth transhumanists are looking for.

As with every breakthrough in medicine, science, and technology, it will be reserved for the military or the wealthy, at least at first. Lt. Col. Robert Maginnis and journalist Annie Jacobsen have explored the potential military uses of cutting edge tech in their books *Future War* and *The Pentagon's Brain*. If there are perceived advantages to restricting radical life extension technologies for the military, you can bet civilians won't see them.

If and when the private sector does get access to medical miracles, the wealthy will be the beneficiaries. You might ask why I'm so skeptical. Fair enough. Consider education: Isn't it in the best interests of society to provide the highest quality education for every child? In fact, wouldn't society be better off if the lowest economic classes were given priority access to the best schools?

Of course it would. So, why do the wealthy send their children to expensive, top-notch learning academies that are out of reach for the poor and middle class?

Because they can. Be honest: If you could do that for your kids and grandkids, wouldn't you?

Now, ask a transhumanist: Why do they think physical and mental upgrades that offer godlike power and near-immortality will be available for everybody?

Next question: Who gets to upgrade? Refer back to the attitude of Zoltan Istvan's fictional transhumanist hero, Jethro Knights:

> Not the losers of the world, or the mediocre, or the downtrodden, or the fearful. They will only drag us down, like they already have.[226]

Obvious follow-up: Who decides who's a loser, or mediocre, or downtrodden? Transhumanism's promise of immortality could easily bring back the eugenics programs of the last century, which sought to improve mankind by removing the unfit from the gene pool.

It is a sad legacy of the United States that the eugenics program adopted by Hitler's Germany was modeled in part on the one in California.[227] Indiana was the first state in the Union to adopt a compulsory sterilization law (1909), but even though eugenics lost public support after World War II because of the horror of the Nazis' implementation, women were still being sterilized in some states until the 1970s.[228]

Although the United States Congress recently repealed the individual health insurance mandate, American progressives will keep fighting for a single-payer system. If they get their wish, how long before the right to have children is restricted for people with a family history of poor health to keep health care costs down for everyone else? And then how long before government requires genetic "improvements" to ensure all babies are healthy? And for a small upcharge, parents can specify height, minimum IQ, gender, eye color, and sexual preference. (Think that's an exaggeration? A National Academy of Sciences panel has already endorsed the concept of "designer babies," at least in limited circumstances.)[229]

We won't even bring up death panels for the elderly. I will, however, point out that a global push to legalize euthanasia is growing. In Belgium, a child who isn't terminally ill may be put to death if a doctor deems them mature enough to make that decision. In the Netherlands, where infanticide is still illegal, doctors admit that killing sick and disabled babies goes on anyway.[230]

Already, transhumanists—unwittingly, I'm guessing, because I can't believe they'd do this on purpose—have adopted the motto of the International Eugenics Congresses held in 1912, 1921, and 1932: "Self-directed evolution."[231]

How quickly we forget.

Next question: Assuming that science somehow overcomes the problems of death and distribution and rolls out the miracle of immortality to everyone on earth, does that really look like heaven?

Here's why I ask: Imagine a world where most of the people, who do *not* subscribe to a biblical standard of morality, are immortal, and therefore no longer restrained from doing things that might have gotten them killed previously. Put another way, imagine a world in which Adolf Hitler, Josef Stalin, and Charles Manson could never die—and where others, who might have been too afraid of dying to live out their vile fantasies, no longer have that check on their behavior.

I don't know about you, but that sounds more like hell than heaven to me.

Last question: How can Christians with a true understanding of what's at stake call themselves transhumanists?

Now the speculation: For several years now, we've considered the possibility that an autonomous artificial intelligence might be used by the Antichrist to give life to the image of the Beast. While this sounds like science fiction, it's not a new idea.

More than a hundred years ago, theologian E. W. Bullinger put forward the very same idea in his commentary on the book of Revelation:

> Nikola Tesla, the Hungarian-American electrician, boldly declares (in *The Century* magazine for June, 1900), that he has a plan for the construction of an automaton which shall have its "own mind," and be able, "independent of any operator, to perform a great variety of acts and operations as if it had intelligence." He speaks of it, not as a miracle, of course, but only as an invention which he "has now perfected."
>
> But again we say we care not how it is going to be done. God's word declares that it will be done, and we believe it… We already hear of talking machines; with "a little" Satanic power thrown in, it will be a miracle very easily worked.[232]

Now, Bullinger was not without controversial beliefs. He was an ultradispensationalist,[233] a believer in the cessation of the soul between death and resurrection,[234] and a member of the Universal Zetetic Society—a flat-earther.[235] Regardless, Bullinger's 1903 observation about the potential prophetic application of Tesla's research was profoundly insightful and relevant to our discussion of the transhumanist movement.

The human body in general, and the brain especially, is a bio-electrical machine. An electroencephalogram measures electrical activity in the brain to diagnose disorders such as epilepsy.[236] As Christians, we know (or we should) that this bio-electrical device can be overwhelmed and controlled by an external entity—it's what we call "demonic possession." Is it possible, then, that an autonomous, electrical superintelligence could provide a substrate for "the image of the Beast"?

Speculative, yes. Has this phenomenon ever been observed? The abomination that was Windows ME notwithstanding, no, not as far as we know. But then, there wouldn't be any advantage for the Enemy to disclose this ability. To quote the famous line by Baudelaire (paraphrased in the popular 1995 film *The Usual Suspects*), "The finest trick of the devil is to persuade you that he does not exist."

This is precisely the type of blind spot to which Christian transhumanists have fallen victim. They're not alone, of course; about 60 percent of American Christians believe Satan "is not a living being but is a symbol of evil."[237] In my forthcoming book *Last Clash of the Titans*, I'll make the case that the horde of Magog in Ezekiel 38 and 39, which is the army that comes against Jerusalem at Armageddon, is linked to the "shades"—the Rephaim, which, I argue, the ancient Amorites believed were the spirits of the Nephilim.

We can't say with certainty that artificial intelligence will produce the false miracle that is the resurrected Beast or an army of demonically possessed super soldiers, but it *is* a fact that leading transhumanists see their goal as transcendence—rising above the limits of our flawed (they think) biology. Inventor Ray Kurzweil, Google's director of engineering, foresees what he calls the Singularity, "a future period during which the pace of technological change will be so rapid, its impact so deep, that human life will be irreversibly transformed."[238]

Kurzweil and his followers mean that literally. They foresee what Dr. Kurzweil calls the "Sixth Epoch of Evolution." In his view, we're now in the final stages of Epoch 4 (humans working with technology) and about to enter Epoch 5, where biology and technology merge to create higher forms of life. Epoch 6 is when "the Universe wakes up," and virtually immortal human-machine hybrids go forth into the universe, presumably to be fruitful and multiply.[239]

Seriously.

As speculative as it sounds, this, like a demonically possessed AI, is not a new idea. The French Jesuit Pierre Teilhard de Chardin, a paleontologist by training and philosopher by nature, believed in the reverse entropy of Darwinian evolution. In his 1959 book *The Phenomenon of Man*, Teilhard theorized that creation was evolving to ever-higher levels of complexity toward something he called the Omega Point. At some future date, he believed, a sphere of sentient thought surrounding the earth he called the noosphere joins with itself, human thought unifies, and "our ancient itch to flee this woeful orb will finally be satisfied as the immense expanse of cosmic matter collapses like some mathematician's hypercube into absolute spirit."[240]

In other words, mind merges with matter and the universe wakes up. Teilhard's Omega Point and Kurzweil's Singularity are the same thing.

Although Teilhard's writings were cited with a warning by the Vatican in 1962, the Pontifical Council for Culture recently approved a petition asking Pope Francis to remove it. The council expressed its desire for the pope to "acknowledge the genuine effort of the pious Jesuit to reconcile the scientific vision of the universe with Christian eschatology."[241] The problem is that any such reconciliation of science and the Bible is neither scientific nor biblical. There is no evidence to support Teilhard's theory of a noosphere as a "living thinking machine with enormous physical powers,"[242] and believing in his Omega Point (and likewise the transhumanist Singularity) requires throwing out the book of Revelation, for a start, and then deleting every other end-times prophecy in the rest of the Bible.

But transhumanists have begun to recognize that appealing to Teilhard's work can help them win over skeptical Christians by providing Christianized language to describe their vision of the future.[243] Christians should see through this ruse. Trading God's promise of a resurrected, incorruptible body for the transhumanists' promise of eternal life for you in a cosmic mainframe is like Esau trading his inheritance to Jacob for a bowl of beans.

Transhumanists believe the Singularity will be humanity's crowning achievement, our great evolutionary leap forward to finally exceed the limits of our biology—in other words, apotheosis, finally realizing the promise from the garden to "be as gods."

This is a sad delusion. Transhumanism is nothing more than the ill-fated quest of Gilgamesh with a sci-fi veneer.

Now, the search will unquestionably yield benefits. We in The Milieu are not technophobes.[244] Medical advances are a good thing, but they are *restoration*, not *transformation*. Christians should never confuse the two. An artificial knee is not the first installment in a full-body immortality upgrade.

According to the epic, when Gilgamesh died, he was laid to rest in a tomb of stone in the bed of the Euphrates River. There is evidence that Gilgamesh didn't go to his eternal rest alone:

> His beloved wife, his beloved children, his beloved favorite and junior wife, his beloved musician, cup-bearer and...., his beloved barber, his beloved...., his beloved palace retainers and servants and his beloved objects were laid down in their places as if.... in the purified (?) palace in the middle of Uruk.[245]

Scholars have debated the meaning of that text for the last hundred years, but tombs of the wealthy at the Sumerian city of Ur, dated at least five hundred years after the probable time of Gilgamesh, included as many as sixty-five servants and retainers.[246] It's possible that this was a tradition that extended back to the kingdom of Uruk ruled by Gilgamesh, and by Nimrod before him.

Sadly, Christian transhumanists are following in the footsteps of Gilgamesh. While we take them at their word that they have accepted Christ as Lord, and are thus ensured of eternity in our Father's house, they, like the king of old, may be leading their beloved spouses, children, friends, and colleagues into death by signaling that Kurzweil is an acceptable substitute for Jesus Christ.

And the journey of unsaved transhumanists will not end inside an earthly tomb. If their trust is in science instead of Christ—in hoping to find their names written in lines of incorruptible computer code rather than in the Book of Life—then their final destination is the second death and a place in the Lake of Fire.

In a section titled "From Antichrist's Mark to His Transhuman Church" from the best-selling book *Forbidden Gates,* Thomas and Nita Horn added these related insights:

> Although most transhumanists, especially early on, were secular atheists and would have had little resemblance to prototypical "people of faith," in the last few years, the exclusion of supernaturalism in favor of rational empiricism has softened as the movement's exponential popularity has swelled to include a growing number of Gnostic Christians, Buddhists, Mormons, Islam, Raelianism, and other religious traditions among its devotees. From among these groups, new tentative "churches" arose—the Church of Virus, the Society for Universal Immortalism, Transtopianism, the Church of Mez, the Society for Venturism, the Church of the Fulfillment, Singularitarianism, and others. Today, with somewhere between 25–30 percent of transhumanists considering themselves religious, these separate sects or early "denominations" within transhumanism are coalescing their various religious worldviews around generally fixed creeds

involving *spiritual transcendence* as a result of human enhancement. Leaders within the movement, whom we refer to here as *transevangelists,* have been providing religion-specific lectures during conferences to guide these disciples toward a collective (hive) understanding of the mystical compatibility between faith and transhumanism. At Trinity College in Toronto, Canada, for instance, transhumanist Peter Addy lectured on the fantastic "Mutant Religious Impulses of the Future" during the Faith, Transhumanism, and Hope symposium. At the same meeting, Prof. Mark Walker spoke on "Becoming Godlike," James Hughes offered "Buddhism and Transhumanism: The Technologies of Self-Perfection," Michael LaTorra gave a "Trans-Spirit" speech, nanotechnologist and lay Catholic Tihamer Toth-Fejel presented "Is Catholic Transhumanism Possible?" and Nick Bostrom spoke on "Transhumanism and Religion."

Recently, the *New York Times* picked up this meme (contagious idea) in its June 11, 2010, feature titled "Merely Human? That's So Yesterday," speaking of transhumanism and the Singularity as offering "a modern-day, quasi-religious answer to the Fountain of Youth by affirming the notion that, yes indeed, humans—or at least something derived from them—can have it all."[247] In commenting on the *Times* article at his blog, one of our favorite writers, bioethicist Wesley J. Smith, observed the following:

Here's an interesting irony: Most transhumanists are materialists. But they desire eternal life as much as the religionists that so many materialists disdain. So they invent a material substitute that offers the benefits of faith, without the burden of sin, as they forge a new eschatology that allows them to maintain their über-rationalist credentials as they try to escape the nihilistic despair that raw materialism often engenders. So they tout a corporeal New Jerusalem and prophesy the coming of the Singularity—

roughly equivalent of the Second Coming for Christians—that will…begin a New Age of peace, harmony, and eternal life right here on Terra firma.[248]

In the peer-reviewed *Journal of Evolution and Technology* published by the Institute for Ethics and Emerging Technologies (founded in 2004 by transhumansists Nick Bostrom and James Hughes), the "Apologia for Transhumanist Religion" by Prof. Gregory Jordan lists the many ways transhumanism is emerging as either a new form of religion or a mirror of fundamental human ambitions, desires, longings, shared hopes, and dreams that traditional religions hold in common. In spite of denial by some of its advocates, Jordan concludes that transhumanism may be considered a rising religion because of its numerous parallels to religious themes and values involving godlike beings, the plan for eternal life, the religious sense of awe surrounding its promises, symbolic rituals among its members, an inspirational worldview based on faith, and technology that promises to heal the wounded, restore sight to the blind, and give hearing back to the deaf.

Of the technological Singularity in particular, Jordan writes how some transhumanists especially view the Singularity as a religious event, "a time when human consciousness will expand beyond itself and throughout the universe." Quoting Kurzweil's *The Singularity Is Near: When Humans Transcend Biology*, Jordan provides:

The matter and energy in our vicinity will become infused with the intelligence, knowledge, creativity, beauty, and emotional intelligence (the ability to love, for example) of our human-machine civilization. Our civilization will expand outward, turning all the dumb matter [normal humans] and energy we encounter into sublimely intelligent—transcendent—matter and energy. So in a sense, we can say that the Singularity will ultimately infuse the world with spirit.

According to these Singularitarians, this expansion of consciousness after the Singularity will also be an approach to the divine:

> Evolution moves toward greater complexity, greater elegance, greater knowledge, greater intelligence, greater beauty, greater creativity, and greater levels of subtle attributes such as love. In every monotheistic tradition God is likewise described as all of these qualities, only without any limitation: infinite knowledge, infinite intelligence, infinite beauty, infinite creativity, infinite love, and so on.... So evolution moves inexorably toward this conception of God.... We can regard, therefore, the freeing of our thinking from the severe limitations of its biological form to be an essentially spiritual undertaking.[249]

Yet while development of a *new* universalist religion appears to be forming among members of transhumanism's enlightenment, conservative scholars will taste the *ancient* origin of its heresy as the incarnation of gnosticism and its disdain for the human body as basically an evil design that is far inferior to what we can make it. "Despite all their rhetoric about enhancing the performance of bodily functions," says Brent Waters, director of the Jerre L. and Mary Joy Stead Center for Ethics and Values, "the posthuman project is nevertheless driven by a hatred and loathing of the body."[250] Transhumanist Prof. Kevin Warwick put it this way: "I was born human. But this was an accident of fate—a condition merely of time and place."

Conversely, in Judeo-Christian faith, the human body is not an ill-designed "meat sack," as transhumans so often deride. We were made in God's image to be temples of His Holy Spirit. The incarnation of God in the person of Jesus Christ and His bodily resurrection are the centerpieces of the Gospel and attest to this magnificent fact. While in our fallen condition human suffering is reality, most traditional Christians believe this struggle makes us stronger and that healing and improvements to the human condition are also to be desired. Throughout history, the Church has therefore been at the forefront of disease treatment discovery, institutions for health care, hospitals, and other medical schools and research centers. In other words, we do not champion a philosophy toward techno-dystopianism. *Indeed, what a day it will be when cancer is cured and we all shout "Hallelujah!"*

But in the soulless posthuman, where DNA is recombined in mockery of the Creator and no man is made in God's image, "there are no essential differences, or absolute demarcations, between bodily existence and computer simulation, cybernetic mechanism and biological organism, robot technology and human goals," says Katherine Hayles, professor of English at the University of California, in her book *How We Became Posthuman.* "Humans can either go gently into that good night, joining the dinosaurs as a species that once ruled the earth but is now obsolete," she says in transhuman contempt of—or outright hostility to—intrinsic human dignity, "or hang on for a while longer by becoming machines themselves. In either case…the age of the human is drawing to a close."[251]

Thus the gauntlet is thrown down and a holy war declared by the new and ungodly apostles of a transhuman faith! We who were created in His image will either adapt and be assimilated to posthuman, or be replaced by Nephilim 2.0 and the revival of their ancient mystery religion. This solidifies how, the more one probes into the ramifications of merging unnatural creations and non-biological inventions according to the transhumanist scheme of seamlessly recalibrating humanity, a deeper malaise emerges, one that suggests those startling "parallels" between modern technology and ancient Watchers activity may be no coincidence at all—that, in fact, a dark conspiracy is truly unfolding as it did "in the days of Noah."

Consider, in conclusion of this chapter, the thoughtful commentary by Dr. C. Christopher Hook:

There are several key questions that our churches and theologians will have to address. Is it appropriate for members of the body of Christ to engage in alterations that go beyond therapy and are irreversible? Is it just to do so in a world already deeply marked by inequities? What does it mean that our Lord healed and restored in His ministry—never enhanced? Is it significant that the gifts of the Holy Spirit—wisdom, love, patience, kindness—cannot be manufactured by technology?[252, 253]

Embedding Transhumanism into Church and Society Through Predictive Programming

By Milieu Member Wes Faull

Transhumanism is the belief or theory that the human race can evolve beyond its current physical and mental limitations, especially by means of science and technology. This is a big idea. The evolution of man. At face value, H+, or Human plus, appears to have many benefits that can help humanity. It has the potential to fight against age and decay, to make our lives easier, and to better equip us to keep up with an ever-changing world. Transhumanism also poses numerous obvious moral and ethical issues that must be addressed before entering this covenant. But the question remains: How would the world ever accept the ideology of transhumanism? With all the unknowns that exist, how could mankind reach a point of accepting a union to merge with monsters and machines? A scheme that is currently in place to achieve this goal is predictive programming.

Predictive programming uses media to subtly influence the viewer into the acceptance of an idea over time. It is the underlying message of a given means of communication. On the surface is a visual montage, or arrangement of images in a sequence that is used to convey a particular message, and underneath lies an indirect suggestion with the intent of eventual implementation in society.

We are living in a time when marketing geniuses, script writers, and producers deserve the applause of Edward Bernays and Joseph Goebbels. Society is constantly inundated by carefully crafted advertisements specifically targeting their audiences. Data mining is at the top of its game, and Google and Facebook are topics we will save for another day. Television shows, or regularly scheduled programming, glue viewers to their screens for an allotted time period and deliver direct and underlying messages to the audience, only to be briefly interrupted by a short break of telling viewers what they need to buy before returning to the conditioning process. In dramas or narratives, viewers relate to a protagonist or antagonist, whatever the case may be, and the content providers can string them along to digest a given ideology almost undetected because of their connection. In reality shows, the viewers are conditioned to accept a standard due to their celebrity worship or their admiration of a particular host. Sports programming plays a similar card—and, as we have seen in days of late, political views and patriotism (or better yet a lack thereof) have crept into a previously unknown arena. The list goes on, but needless to say, television communication plays a huge role in how we view and process the world. The same can be said of the movie industry. With this in mind, here is an example of predictive programming. Insert your favorite suggestion here.

At the beginning of the process, the suggestion might be placed in a fictional narrative or even science-fiction format in a movie. By doing this, the viewers can be taken farther down the path because the idea isn't presented in the realm of reality or as actually ever occurring in society. Then the idea reappears in a few different television shows—the more episodes the better. To further embed the suggestion, more movies are made to present the topic in a positive and negative light; this way, the topic opens up for discussion and the desired outcome is publicized. This is followed by disbursing the theme into both print and digital media advertisements. Wash, rinse, and repeat. During each cycle, a higher frequency of the suggestion is added, and more reality is incorporated to keep up with current trends. The familiar analogy of the frog in the pot of water comes to mind: If you drop a frog in a boiling pot of water, it will immediately remove itself from said pot, but if you slowly turn up the temperature, the frog will remain in the pot because the change in temperature is seemingly undetectable. The downside is that regardless of the rate of escalation, boiling water is an unfortunate ending for the targeted frog, as is transhumanism for the world. Predictive programming slowly desensitizes and conditions the public, numbing it into submission and acceptance. Before I go too far in this direction, let me tie this back to my personal life and the idea of transhumanism.

I am a child of the late '70s and '80s. Among my favorite cartoons growing up were DC Comics- and Marvel-based programming, namely, the *Super Friends* and *Spiderman*. Close seconds were the *Fantastic Four* and *Captain America*. I spent many a day watching the Justice League battle the Legion of Doom, as well as seeing Spiderman swing in to come and save the day. I clearly remember the desire to have superpowers I could use to battle evil and fight injustice in the world. A closer look today shows that my childhood heroes were actually the products of transhumanism. Peter Parker's DNA was altered from a bite by a radioactive spider. He became Spiderman when he gained the abilities of superhuman strength, Spidey sense, agility, superb acrobatic skills, spider web glands, and the capacity to stick to most surfaces.[254] Barry Allen's DNA was altered when a lightning bolt struck his lab and rendered him The Flash. The metahuman speed he received from his electrocution provided him with the means to defy gravity, run up walls and over water, and move faster than time itself.[255] The Fantastic Four DNA alteration was a result of their being accosted by cosmic rays while they were on a trip into outer space. After the change, Reed Richards (Mister Fantastic) could stretch his physique and skeletal structure beyond the measure of human understanding. Susan Storm (Invisible Girl) gained the ability to cloak herself with invisibility by bending light around her flesh. Johnny Storm (Human Torch) gained flight and fire. Benjamin

Grimm (the Thing) received superhuman power and rocklike stability from the blast; regrettably, it also caused him to suffer becoming very unpleasant to look upon.[256] Steve Rogers participated in Project: Rebirth, a secretive governmental project led by a pre-DARPA organization if you would, that cured his polio, gave him superhuman strength, and brought Captain America to the world.[257] I went on journeys with these heroes for years, and appreciated how they used their special powers for the good of humanity. All of these characters began as mere mortals, but became as gods. They also seemed to share a good upbringing, hold the same morals and values, and use their newfound abilities for the good of mankind. How could anything be wrong with becoming a transgenic superhuman?

On the demigod side of the coin, my favorites were Superman, Wonder Woman, and Aquaman. These beings were born this way. Superman (Clark Kent), or Kal-El (Voice of God) was born on the planet Krypton, and was endowed with an arsenal of superpowers. He could fly, and he had heat vision, freezing breath, and superhuman strength and speed. He used these abilities to protect those who dwell on the earth.[258] Wonder Woman (Diana Prince) was born on the island of Themyscira and was the offspring of Queen Hippolyta of the Amazons and Zeus. Her extraordinary agility, strength, endurance, heightened senses, combat skills, and the ability to quickly heal during battle were all genetically upgraded gifts from the Greek gods.[259] Similarly, Aquaman (Arthur Curry) was the product of a love affair between Atlanna, Queen of Atlantis, and a lighthouse keeper from Maine named Tom Curry. He was another who had superhuman strength and resilience, along with the abilities to telepathically control sea life, swim really fast, and breathe underwater.[260] All of these were born with posthuman abilities they used to keep humanity safe. They, too, seemed to have been raised by parents with high moral standards that were successfully passed down to them and that they continued to display in their everyday lives. Could being born with advancements or even manipulating the human genome to achieve this heightened level of existence ever have negative repercussions?

Enter the Legion of Doom. Assembled and led by the mastermind of Lex Luthor, this transhuman group of supervillains is a prime example proving that evil exists and surpasses all levels of technology, science, and societal progression. Evil is an issue of the heart. Of the extraterrestrial supervillains, Bizarro was from the planet htraE, which is "earth" spelled backwards, and he was a well-known archenemy of Superman. He had all of the same superhuman abilities as Superman, yet he chose to use them to hinder those who fought for truth and justice.[261] Brainiac is another being from a different galaxy who was on a mission to attain the knowledge of the universe. He is known to have taken civilizations captive and bottle them as trophies from his quest, while leaving their planets in ruin.[262] Sinestro from the planet Korugar is a great example of the moral question. He was one of the greatest Green Lanterns and excelled at keeping his jurisdiction of the planet at peace, but pride and power caused him to defect from the corps and take his tyrannical mission of "peacekeeping" to the universe at all costs.[263] There were also members of the gang who were born mortals and became supervillains. Giganta, for example, was transformed due to magical dust stolen from Apache Chief, and it caused her to grow. Since the nature of her heart was evil, she not only grew in strength and stature, but also in iniquity. She was in like company with the Legion of Doom and used her ability to thwart the efforts of the Super Friends.[264] Cyrus Gold was born in Gotham City and lived a nefarious life. After his death in Slaughter Swamp, he was resurrected with immortality, superhuman strength, and a quality of being nearly impervious, becoming Solomon Grundy.[265] Just because someone is born posthuman, or even if he or she becomes Human 2.0, we cannot presume that the person will automatically possess perfected behavioral qualities.

Even though these are children's cartoons, this analysis should help solidify how predictive programming works. The underlying messages delivered by the medium of these cartoons were all steeped in suggestion of the idea of transhumanism. These fantastic fictional stories—cartoons, mind you—all planted a posthuman, superhuman desire in my innocent mind. I wasn't much of a comic book reader when I was young, but that was yet another medium of delivery. These stories were also communicated in various television live-action programs as well. Fast forward to today and we find a revamping or new cycle of the same. The new cartoon presentations of these characters and others have continued to entertain the next generation, and their live-action television shows are doing well. Currently, the CW Network broadcasts the *Green Arrow, The Flash*, the *Legends of Tomorrow*, and *Supergirl*. Fox keeps up with the competition via *Gotham, Legion*, and its new addition, *The Gifted*. Netflix productions include *Daredevil, Jessica Jones, Luke Cage, Iron Fist*, and *The Punisher*—all of which feature either protagonist and/or antagonist transhumans. Comic-book stories predate my birth by decades, and I don't know them well enough to give a comparison and contrast of their content versus their transformation into animation, television shows, and films, but I do know that, today, the final product as far as production value is truly amazing! Hence, the current revival of DC and Marvel on the big screen. Box-office smashes that deserve naming include *The Avengers, Batman, Superman, Wonder Woman, Spiderman, Guardians of the Galaxy, Justice League*, and *X-Men*. The delivery of these epic tales in 4K high definition easily trumps that of the black-and-white days of *Superman* on TV, *The Flash* of the '80s, or even Christopher Reeve flying in front of a green screen. The modern-day viewer can now voluntarily be thrown into the presentation via 3D and 4D IMAX experiences and witness the battles firsthand, plugging themselves into the hardwired conditioning. As the desire grows for superpowers and human evolution from the updated theater experience and television series to boot, video games accompany the effect with first-person controls. Salt that with today's propaganda on accepting the homosexual agenda, political stabs, and the divide-and-conquer mentality on all fronts, and it's obvious that we are being attacked by the best programmers money can buy.

On a different but similar note is the topic of getting the Church to accept transhumanism. The Church is supposed to be "in the world but not of it." Unfortunately, it is harder to decipher between the two today than it used to be. Regardless, the Church is in the world, so we, too, are affected by the aforementioned programming techniques in media. Outside of this dilemma, we also have to be on our toes in regard to doctrine being taught from within. To the true Body of Christ, the Church, His Bride, the Bible is our map for existence. It is our basis of truth. It is the Truth. It is the written Word of God. From it we find that Scripture tells us to "test every spirit." The Word is "God-breathed" and used for "doctrine, reproof, and correction to establish us for every perfect work." This standard is what we test everything by in order to know what is truth, because there are some who will pervert it. We find many observations in the Bible that directly apply to transhumanism or becoming posthuman.

We read in Philippians 3:20–21 that:

Our conversation is in heaven; from whence also we look for the Saviour, the Lord Jesus Christ: Who shall change our vile body, that it may be fashioned like unto his glorious body, according to the working whereby he is able even to subdue all things unto himself.

And Paul, speaking further in 1 Corinthians 15:42–54 of the human body and the resurrection of the dead, says:

It is sown in corruption; it is raised in incorruption: It is sown in dishonour; it is raised in glory: it is sown in weakness; it is raised in power: It is sown a natural body; it is raised a spiritual body. There is a natural body, and there is a spiritual body…. And as we have borne the image of the earthy, we shall also bear the image of the heavenly. Now this I say, brethren, that flesh and blood cannot inherit the kingdom of God; neither doth corruption inherit incorruption. Behold, I shew you a mystery; We shall not all sleep, but we shall all be changed, In a moment, in the twinkling of an eye, at the last trump: for the trumpet shall sound, and the dead shall be raised incorruptible, and we shall be changed. For this corruptible must put on incorruption, and this mortal must put on immortality. So when this corruptible shall have put on incorruption, and this mortal shall have put on immortality, then shall be brought to pass the saying that is written, Death is swallowed up in victory.

John hits the same topic when he tells us in his first epistle:

Beloved, now are we the sons of God, and it doth not yet appear what we shall be: but we know that, when he shall appear, we shall be like him; for we shall see him as he is. (1 John 3:2)

There is a lot of information to process, so let me summarize the passages addressing believers in Christ. Jesus Christ will return from heaven and change our vile bodies into glorified bodies and subdue all things to Himself. The current fallen flesh we occupy is sown in corruption, dishonor, and weakness, and will be raised in incorruption, glory, and power. The natural earthly body can't inherit the kingdom of God, and it will be converted into a spiritual body in order to do so. The mortal must put on immortality when it is changed and death will be swallowed up in victory. The biggest thing not to miss here is that our miraculous changing into immortality, becoming completed in Christ, or posthuman, is achieved through none other than Jesus, the Christ. It happens when we see Him as He is, when He "himself shall descend from heaven with a shout, with the voice of the archangel, and with the trump of God: and the dead in Christ shall rise first: Then we which are alive and remain shall be caught up together with them in the clouds, to meet the Lord in the air: and so shall we ever be with the Lord" (1 Thessalonians 4:16–17).

All these verses confirm the same idea. As human beings, each of us is a soul with a body and not a body with a soul. The soul is our being, the inner man, and currently it exists housed within a flesh shell. This allows us to operate in this realm on earth and live our lives. Because of sin entering the world through man, causing us to be separated from God, these flesh-clothed bodies have been given an expiration date; they have been slowly dying since we began our habitation of them. If sin didn't enter the world through man, then mankind would still be in a physical and spiritual, face-to-face existence with God; we would be immortal; and there wouldn't have been a need to prepare another path to regain access to the Tree of Life. The Garden of Eden would still be open for business, and we wouldn't have to deal with sickness, disease, and death. Thus, is the Fall of Man. This is the reason we are all in need of a Savior. That is why the redemptive work of Jesus Christ at the cross is necessary and the only qualification that we have. This is the hope we have in Christ. He is sovereign, He is faithful, and He does what He says He is going to do. Count on it! The transformation that Scripture tells us will happen specifically revolves around and is completely dependent upon Jesus the Messiah. Only because of what He did and the mission He accomplished on earth allow us to one day be fully completed in Him. Transhumanism is man's effort of subverting the cross and an attempt to physically and spiritually achieve immortality without Christ. There are resurrection bodies that the elect will put on when we are changed and see Him as He truly is, but Jesus the Messiah will be the author and finisher of that transaction. He is the only one with the authority and the ability to do so. Humanity won't prevent this event, and transhumanism won't be the culprit. What if someone told you the Bible says you can achieve this prior to His return?

Manifest Sons of God

Unfortunately, as previously mentioned, there are some who pervert God's Word. An idea has resurfaced today called the "manifest sons of God" doctrine. This ideology is pulled from Romans 8:19: "For the earnest expectation of the creature waiteth for the manifestation of the sons of God." It also pulls from 1 John 3:2, and is the idea that mankind can attain immortality or completed state before Christ returns and without the need for Him to be completed. It holds hands with the dominion-istic Kingdom Now theology whose adherents believe Christ will not return until the Church prepares the world for Him. Further, we can't successfully complete this task for Christ until the manifestation of the sons of God occurs. By means of "the Holy Spirit" we can attain the status of manifest sons of God, become immortal, and help usher in the kingdom of God to assist in bringing down the Antichrist. Accord-ing to one author/teacher, "The Earth and all of creation is waiting for the manifestation of the sons of God, the time when they will come into their maturity and immortalization.… When the Church realizes its full sonship, its bodily redemption will cause a redemptive chain reac-tion throughout all of creation."[266] Remember that this redemption and immortalization are not dependent on the Church coming to a realiza-tion; the Bible tells us clearly that it is solely dependent on the Lord Jesus Christ.

Kenosis Christ

To further complicate the issue, I also need to add that the "Jesus" being taught by this belief system is a kenosis Christ. In Philippians 2:7, Paul says that Jesus "made himself of no reputation, and took upon him the form of a servant, and was made in the likeness of men." According to R. C. Sproul, this means that Jesus "willingly cloaked his glory under the veil of this human nature that he took upon himself."[267] In other words, the inner being or soul that inhabited the flesh shell of Jesus of Nazareth for 33 years was 100 percent God. Contrary to this understanding is kenosis. Simply put, Kenostics believe that the making "himself of no reputation"—or "emptying of himself," as some translations put it—is interpreted to mean that Jesus laid aside His divinity at incarnation and lived life as a man led by the Holy Spirit. Therefore, we can do the same and become the manifest sons of God. The idea holds that Jesus was the first of the manifest sons of God, and we are commanded to walk in His footsteps and go and do likewise. This opens the door for transhumanism to invade church doctrine.

But I believe Jesus is God. Jesus always has been God. Jesus always will be God. He is the same yesterday, today, and forever. Kenosis reduces Jesus to a man who was sinless and in right standing before God, which allowed Him to work the supernatural while on earth. By lowering Jesus to man's level, Kenosists offer that His level is attainable by man. Supernatural abilities, miracles, that normal men aren't born with are also attainable. Become one of the manifest sons of God and you, too, can perform these signs and wonders. Transhumanism would be a fast track to these ends.

Joel's Transhuman Army

Joel's Army is a similar doctrine that holds hands with this train of thought. It follows Kingdom Now theology in the belief that the Church was given dominion over the earth in Genesis 1:28, and once we regain this authority, it will be ready for Christ to return. A big problem with this idea is that God granted Adam dominion over the animals, not over humanity. Also, Adam and the Church are not interchangeable. It is not up to us to subdue mankind into submission for Christ to return. It is our command to go and make disciples of every nation. Only the Holy Spirit can draw a person to God. Back to Joel's Army. This doctrine pulls from a passage located in Joel chapter 2, which is an exciting read. When reading the descriptions of this demonic army, it almost sounds like Lex Luthor plotted with a physicist and produced an apocalyptic Legion of Doom to fight against the Justice League. This takes place on the Day of the Lord.

[2] A day of darkness and of gloominess, a day of clouds and of thick darkness, as the morning spread upon the mountains: a great people and a strong; there hath not been ever the like, neither shall be any more after it, even to the years of many generations.

[3] A fire devoureth before them; and behind them a flame burneth: the land is as the garden of Eden before them, and behind them a desolate wilderness; yea, and nothing shall escape them.

[4] The appearance of them is as the appearance of horses; and as horsemen, so shall they run.

[5] Like the noise of chariots on the tops of mountains shall they leap, like the noise of a flame of fire that devoureth the stubble, as a strong people set in battle array.

[6] Before their face the people shall be much pained: all faces shall gather blackness.

[7] They shall run like mighty men; they shall climb the wall like men of war; and they shall march every one on his ways, and they shall not break their ranks:

[8] Neither shall one thrust another; they shall walk every one in his path: and when they fall upon the sword, they shall not be wounded.

[9] They shall run to and fro in the city; they shall run upon the wall, they shall climb up upon the houses; they shall enter in at the windows like a thief.

[10] The earth shall quake before them; the heavens shall tremble: the sun and the moon shall be dark, and the stars shall withdraw their shining:

[11] And the LORD shall utter his voice before his army: for his camp is very great: for he is strong that executeth his word: for the day of the LORD is great and very terrible; and who can abide it? (Joel 2:2–11)

The chapter goes on to show that this army is allowed to come in and wreak havoc in order for the people to repent of their sin, fast and pray, and return their hearts to the Lord. If they do this, the Lord will turn away this army from them, send it to a barren land, and deal with them accordingly. Believers in the Joel's Army doctrine have a different understanding of this passage. They hold that the transhuman army described here, very reminiscent of the army from the bottomless pit spoken of by the Apostle John in the book of Revelation, will actually be believers who are given supernatural abilities. They are convinced they will become supernaturally equipped to fight against the Antichrist and release God's judgments on the non-believing world, as well as "normal" Christians who get in their way.

Once again, Scripture tells us the blood of Jesus Christ cleanses us from all sin (1 John 1:7). Salvation is by grace through faith in Jesus, not of works (Ephesians 2:8), and the Church is not the redeeming agent. If this scenario is the desired commission being taught by some in the Church, one would think that using science and technology to achieve this level of existence would not be looked down upon.

One last area to reference within the Church is the Christian Transhumanism Association. I'm sure this idea spans denominational lines, so I won't put a blanket over them to combine them into Dominionism, Kenosticism, Manifest Sons, or Joel's Army followers. Without coupling them to the previous misinterpretations of Scripture desiring to become Human plus before Christ returns, their self-professed title still clearly announces their stance on the topic of becoming posthuman. They see transhumanism as a chance to use "science & technology to participate in the work of God, to cultivate life and renew creation."[268] That mission statement does sound very appealing: participating in God's work, cultivating life, and renewing creation…and using science and technology to do it. The concerns I have with this ideology in relation to creation and God's plan for it are as follows: God is the Creator of all things. Mankind is such a prized possession to Him that, after man separated himself,

Jesus still gave His own life to save it. God is our renewal and He is the Renewer. He makes all things new. Mutating and manipulating DNA, the structure used to build man, isn't a renewing alteration; it is a corruption of what God said was good after He created it. Transhumanism is an abomination in direct defiance to the Lord.

Acceptance goes back to predictive programming. It's much easier to see, feel, and understand how the process works with print, and more so with cartoons, television, and film. But, live presentations, speeches, sermons, and stage plays all predate the history of motion pictures. Predictive programming is an age-old method of directing society. All forms of communication effect the recipient upon receipt, through the delivery of a message. It is vital to pause and process the information before digesting it. This is even more important within the Church. Unfortunately, just because someone claims to be anointed and say that he or she comes in the name of Jesus doesn't mean that person has the Church's best interest at heart. Christ tells us in Matthew 24:4–5 to "Take heed that no man deceive you. For many shall come in my name, saying, I am Christ; and shall deceive many." This is simple, but many times overlooked. This passage doesn't exclusively mean to be aware of people falsely claiming to be Jesus, although that is also covered. Specifically, in Greek, "Christ," or *Christos,* means "anointed"—that is, "the Messiah," an epithet of Jesus. It comes from the root word "anoint," or *chriō,* which means "to smear or rub with oil that is by implication to consecrate to an office or religious service." Jesus warns us to "take heed," *blepō,* the same word used in Revelation 6, to come and see, behold, that we are not deceived by anyone—and there will be many coming in His name and claiming to be anointed or consecrated to an office. Know the truth, and the counterfeits will be become obvious.

Let's take a last moment to draw a connection. Satan is the ultimate counterfeiter. He always has been. He always will be. As we have the Holy Trinity: God the Father, Jesus the Son—the Christ, and the Holy Spirit, Satan has himself playing the role of the Father, the Antichrist is in the place of Jesus, and the False Prophet is in the place of the Holy

Spirit. As we are all one in Christ who are the Church, the Body of Christ, in a bizarro world is the body of the Beast. Our resurrected, completed, immortal bodies in Christ are offered counterfeits via transhumanism's desired end game of evolved H+ with a hive mind, completing the Singularity. Selah.

With the differing doctrines creeping in unaware into churches, added to the fascination of the revived superheroes, it's not a huge leap to see how many could be deceived and fall for the idea of a transhumanist future. The Body of Christ must stand fast! It is late in the game, and there is no other option. Don't be one of those who perishes due to lack of knowledge. Continue to keep the Lord first in your life. Continue to stay in the Word and filter everything in this life through it. Continue to hold one another accountable and edify the Body. Carry one another's burdens and walk in love. Diligently live ready!

From Flawed Man to Enhanced Beast
The Philosophical and Political Dilemmas of Transhumanism

By Milieu Member Josh Peck

Introduction: The Issue of Morality

Like many new technologies and developments in their infancies, transhumanism makes powerful claims to justify the movement as morally correct. Among these are promises of a cure for cancer and every other conceivable disease, vast extension of longevity to the point of immortality, and an evolution of the human species. This view of morality is separate from the issue of human biology, yet it *defines* humanity. For example, in an article from *The Journal of Medicine and Philosophy: A Forum for Bioethics and Philosophy of Medicine* entitled "Moral Transhumanism," authors Ingmar Persson and Julian Savuescu argue the point that humanity is defined by morality rather than biology.[269] To quote:

> This paper argues that biomedical research and therapy should make humans in the biological sense more human in the moral sense, even if they cease to be human in the biological sense. This serves valuable biomedical ends like the promotion

of health and well-being, for if humans do not become more moral, civilization is threatened. It is unimportant that humans remain biologically human, since they do not have moral value in virtue of belonging to H. sapiens.

It would seem, at least to some transhumanists, humanity is fundamentally a moral pursuit rather than a biological fact. Some obvious issues with this interpretation should come to mind immediately. How do we define morality? The *Stanford Encyclopedia of Philosophy* provides two possible definitions for morality:

1—descriptively to refer to certain codes of conduct put forward by a society or a group (such as a religion), or accepted by an individual for her own behavior, or

2—normatively to refer to a code of conduct that, given specified conditions, would be put forward by all rational persons.[270]

If we consider the first definition in light of humanity, would this mean some groups, societies, or (dare I say) races are less human if they have differing moralities? If they are, how could anyone else rationally hold them to any other moral structure? If not, and if there is one, central morality that should be accepted by all humans, how do we define exactly what that moral structure looks like? If it is defined by religion, would this mean some religious beliefs offer greater humanity to some rather than the religious beliefs of others? Is a Catholic more human than a Protestant? Is a Buddhist more human than a Muslim? How would atheists rank on this scale?

The second definition provides even more worrisome problems. Who exactly defines the terms of morality that should be generally accepted? How do we define a "rational person?" If someone disagrees with the status quo of morality due to religious or cultural reasons, does it make that person less human?

While these two definitions outline fairly well what morality actually is, it shows us that "morality" does not mean "humanity." To conflate the subjectiveness of a concept such as morality with the biological fact of humanity is to do a grave disservice to all religions, races, children, the disabled, and countless others. It defines a select few as the most human while leaving the rest of the people on earth on some odd, varying scale of humanity that is impossible to objectively define. The only true, scientific, and objective definition of humanity is based on biology rather than morality.

> A human being can be defined in a broad sense quite simply. The *English Oxford Living Dictionaries* defines "human being" as: "A man, woman, or child of the species Homo sapiens, distinguished from other animals by superior mental development, power of articulate speech, and upright stance."[271]

The majority of transhumanist goals, if successfully carried out, would not fit within this definition, or rather, would surpass it. *Homo sapiens* now, compared to other animals, are considered to have superior mental development. However, one of the goals of transhumanism is to increase that development to a state far surpassing what is possible to mere *Homo sapiens*. We now have the power of articulate speech, yet if something like uploading a person into a global brain is available, the speech available in that environment would no longer be articulate or even understandable to normal *Homo sapiens*. Even the idea of upright stance would be a moot point in a virtual environment.

This is why many transhumanists explain transhumanism as a directed evolution from *Homo sapiens* to something new, improved, enhanced, and wholly different. This type of being is called *Homo superior* (H+), or sometimes *Homo futures*.[272] One of the more troubling things about this concept is how this will be done and what it means for the rest of humanity.

Much of transhumanism relies on the implementation of genetic editing. In a nutshell, the idea is a person can alter his or her genes and/or DNA to enhance that person's overall biology. This would greatly reduce the limitations of human biology. Certain enhancements might include relatively small "upgrades," such as giving people super strength or night vision, but could also be as extreme as immortality or an immunity to every known disease. While these are certainly tantalizing and tempting outcomes, there is no way to know the future ramifications of these enhancements.

One definite and unavoidable consequence to transhumanism is the passing down of edited/enhanced genes through generations. This means that, eventually, a world could conceivably exist where the choice to conceive human children with a human partner would be totally removed, thereby forcing the transhumanism agenda onto the rest of the human population against their will. It might be defined as merely a personal choice today, but that is only remotely valid for, at most, one generation. For example, getting a tattoo is a personal choice, yet when a tattooed person has a baby, the baby is not also born with tattoos. With transhumanism, the future generation has no choice, and, once this choice is made for members of that future generation, there is no going back. A *Homo superior* baby could never grow up and decide to be a *Homo sapien*, even if he or she morally disagrees with what transhumanism has done to his or her biological makeup.

Also, a *Homo superior* is not just created out of thin air; a *Homo sapien* is needed to create a *Homo superior*, which means for the addition of every *Homo superior*, one *Homo sapien* is lost. Over time, these two factors (the shrinking *Homo sapien* gene pool and the loss of existing *Homo sapiens*) would eventually result in a complete species eradication. This is known as *specicide.* While the great evil of genocide is defined as "the deliberate and systematic extermination of a national, racial, political, or cultural group,"[273] specicide (in the context of this chapter) is the actual extinction of the entire human species.

If one accepts the current model of natural evolution, we see that one species is not made extinct to create another, more advanced species. For example, primates were not made extinct by nature to create human beings. One might argue that natural selection contributed to the extinction of species; however, this is not the same context as with humanity versus transhumanism. The pursuit of transhumanism is not natural evolution; rather, it is unnatural interference. The eradication of *Homo sapiens* is not natural selection; rather, it would be more comparable to the unnatural wiping-out of an animal species due to unnecessary human intervention. For example, if North Korea launched a nuclear weapon at the Arctic Circle region of the world, killing all the polar bears, would anyone consider that to be natural selection?

This is why defining transhumanism as moral or even as a step in the pursuit of morality is irrational. Taking what is arguably a natural process into our own hands as human beings, altering it at the expense of an entire species and subjecting others to that decision against their will is, by any stretch of the imagination, immoral.

This is not to say that transhumanists today are immoral or do not have humanity's best interest at heart. Personally, I do not view the stated intentions of transhumanism as evil. However, to truly make sure it is best for humanity, any position, idea, or belief (including religion) should welcome criticism and scrutiny. It is the only way to know for sure if our way of thinking is correct, by holding it logically against opposing views and weighing out the options. I certainly do this as a Christian and invite other Christians to think critically about their beliefs as well. I believe this is what can strengthen our beliefs and allow us to abandon certain traditions or cultural interpretations that do not hold up to scrutiny against the whole of the Bible.

In much the same way, it is beneficial to allow scientific pursuits to be held against logical scrutiny as well. In fact, I would argue that it is foolish not to do so. In doing this with transhumanism, it is my opinion that this idea goes well beyond a personal choice. If it was merely a personal choice, my introduction to this chapter would be far different. However, since it has the potential to affect the rest of us, it is worth acknowledging up front before getting into the other moral, political, and economical ramifications of transhumanism. Regardless of how much it may seem like a personal choice on the surface, the current transhumanist agenda has no other outcome than that of unavoidable force, enslavement, and eventual extinction.

Legislating Morality

One main difficulty in processing the transhumanist agenda is to define what is immoral and what should be illegal. Immorality and illegality are two very different things. Morality varies slightly from person to person, and even more culturally. A person could disagree with what I may believe to be immoral without bringing the law into it whatsoever. For example, I believe adultery is immoral. I would even argue this to be an objective moral issue rather than a subjective one. Adultery breaks up families, emotionally damages the children of a cheating spouse, and has the ability to radically change the worldview of all affected. However, in America, adultery is not illegal. There are many things in our culture the majority of people consider immoral, yet for good reason, they remain legal.

We certainly would not want to live in a culture where every accepted immoral thing was made illegal. If this were the case, our First Amendment right to free speech would be completely eradicated. Lying is immoral, but if it was made illegal, we would no longer live in a free society. When it comes to free speech, either all speech has to be free, or none of it can be.

There are also things that are illegal yet not necessarily immoral. For example, failing to stop at a stop sign is illegal. Now, if running a stop sign is done on purpose, a case could be made that it is immoral, because it potentially puts other people at risk of injury or even death. However, if a stop sign is run accidentally, while possibly careless and stupid, it would not necessarily have to be considered immoral. When it comes to the traffic law itself, however, intent isn't a factor. Either a person runs the stop sign or doesn't. If he or she does, that person has committed an illegal act.

Morality, unlike the law, differs from person to person. A Christian might have a different set of moral guidelines than a Buddhist, though some guidelines for morality may overlap. Western culture by and large has some differences in moral guidelines than Eastern or Middle-Eastern cultures. Therefore, to have civilized society, law and morality are both needed, though they may differ at times.

In American culture, a morality of freedom is in place amongst most people. This morality basically says, "Leave me alone and I'll leave you alone." It is the idea that a person is free to do what he or she wants as long as it is not hurting someone else or stopping others from exercising their freedom to do what they want. One of the main exercises in this type of morality is the freedom for people to do what they wish to their own body, also called "morphological freedom." As long as it is not adversely affecting others, the American cultural understanding of morality allows for it.

This tenet of American morality is what transhumanism stands on. The main argument is, "I should be able to do whatever I want to my body." Yes, the principle is true; the way it is used is a bit disingenuous. If transhumanism was kept specifically to one's own body, then it would be a nonissue in American society. However, as discussed earlier, transhumanism is not confined to those who wish to participate in it. Eventually, as early as within a single generation, altered genes and mutations absolutely will spill over into the gene pool of the rest of humanity. Therefore, it is an issue that affects not only the transhumanist, but also the non-transhumanist who wishes to keep his progeny completely *Homo sapien.*

This causes even further problems for religiously minded people such as Christians. In America, citizens are supposed to have the freedom of religion, meaning they can practice their religion without persecution from the government or other citizens. A major Christian belief is that the physical body is a gift from God, and it is imperative to treat it respectfully in accordance with biblical teaching. Altering genetics into something other than human, in some Christian interpretations, could be considered as forfeiting the image of God after which all humans are created. A Christian would not want to force someone to not do something immoral if it only affected the person doing it, but when a person wants to do something that will affect the rest of humanity in a way that revokes the right to keep an unaltered human progeny, most Christians will find it unacceptable.

The question also comes up: If successful transhumanists in the future are no longer considered human, are they subject to accepted human rights? Would transhumanists require their own Bill of Rights? This is what one transhumanist has proposed.

Zoltan Istvan is a leader in the Transhumanist Movement.[274] He ran for president in 2016 to bring awareness to the transhumanist agenda. He is the author of *The Transhumanist Wager*, a novel delving into the ideals of transhumanism.

In 2015, Zoltan launched a five-month national tour in a bus shaped like a coffin to spread the messages of transhumanism. At the end of the tour, Zoltan delivered the Transhuman Bill of Rights to Congress. Zoltan said:

> You read the United Nations Declaration of Rights, which most democracies base their bill of rights on, [and] here we are 65 years later, there's nothing in it that talks about cyborgism. There's nothing in it about whether you can torture an artificial intelligence or a robot. There's nothing in it about whether you could marry robots. I can tell you transhumanists and cyborgs, people that want to try these things will fight for them.[275]

Among others listed, the Transhuman Bill of Rights includes: a universal right to live indefinitely and eliminate involuntary suffering through science and technology; the belief that growing old should be treated as a disease; and the tenet that, under law, no cultural, ethnic, or religious perspectives influencing government policy can impede life-extension science, the health of the public, or the possible maximum amount of life hours citizens possess.[276]

Capitalist-vs.-Socialist Approaches
to the Transhumanist Agenda

The political and economic views of prominent transhumanists are varied. Some see the advantage in privatizing technological research and development to promote competition in the free market with the hopes of faster and more effective advancement. For example, imagine if Apple was solely developing products for transhumanists. Most likely, the technology needed to complete the goals of the transhumanist agenda would far surpass what is available now.

Others, on the other hand, take a more left-leaning approach and believe it is the role of the government to advance the efforts of transhumanism. While this approach takes the financial responsibility away from private citizens, it gives all the power to the government. This means the government can advance the cause as slowly or as quickly as it wants. Also, this gives regulatory power to the government. If the government holds the patents and rights to these technologies and products, it will be the decision of the government how they are used and who has access to them. This approach, while on the surface seeming to decrease cost to private citizens and companies, has the added disadvantage of an inevitable raise in taxes. After all, how would the government pay to advance transhumanist research and develop new technologies? Either it cuts costs somewhere else, or, far more likely, raise the taxes of all American citizens. This creeps into the area of immorality, as there are people who would not wish to have their tax dollars pay for transhumanist technologies. Yet, it is illegal to refuse to pay taxes.

As of now, it is unclear if the costs of transhumanist technology will be privatized or socialized; however, there are issues with either scenario. If transhumanist technology is privatized, then taxes will not have to be increased to fund it, the technology would likely advance far more quickly, and the actual products—be they microchips, injections, pills, or anything else—will become available at the lowest cost possible to the consumer due to free-market competition. However, this would also result in a lack of regulation, just as it is today. Transhumanist technologies would be free to spread throughout the population, affecting even those who do not wish to participate in a variety of ways. For example, what if a transhumanist with edited genes anonymously donated blood, and a person not wanting to participate in transhumanism received this blood unknowingly? What if a person is not required to disclose that he or she has altered genes before having sex with someone else? Eventually, edited genes could easily become the norm, and those not wishing to be a transhumanist or bring transhumanist offspring into the world will have that choice revoked. As stated earlier, this would mean the end of *Homo sapiens.*

On the other hand, if transhumanism becomes a government-funded program, the government would make all the rules and decide how the technology is used. This could still easily result in altered genes in with the genes of the general population, such as the previous scenario, yet there are other concerns to consider. The idea of the government making all the decisions might not be worrisome if the people running the country were completely trustworthy and honest; however, this has never been the case. Power corrupts, and in this scenario, the government would hold all the power. American citizens, transhumanists and nontranshumanists alike, would have to trust the government to use the technology responsibly, yet many, including myself, would not be enthusiastic or optimistic about that idea. What if they decide to only keep it for themselves? What if they decide to go down the road of eugenics and human and/or transhuman experimentation? If a transhuman is not protected by the Bill of Rights, what would stop the government from testing and experimenting on people it has made into transhumanists? What could begin as an innocent medical trial could easily turn into a horrific and tortuous nightmare.

Imagine if there was a voluntary medical test offering a cash incentive for people taking part in it. Imagine it is to test a cure for the common cold through gene editing. The medical professionals infect you with a cold, then use a process of genetic editing to cure you. However, what if, by the standards and understanding of the culture, this now means you are no longer considered human or *Homo sapien*, but are now considered *Homo superior*, set apart from humanity? Now you find yourself in a situation where you are not protected by any kind of basic human rights. You are now the property of the government to be used however it wishes. Would you trust the government enough to think you will be well taken care of? I certainly would not.

Either approach, whether more capitalistic or socialistic, offers some very troublesome potential outcomes to consider before diving headfirst into the transhumanist agenda. Some of these technologies are already available, yet, so far, no legal standard or moral guidelines within our culture have been put in place to protect the rest of the population or even transhumanists themselves. I, like any other rational person, do not want to see harm come to anyone, whether transhumanist or not.

Conclusion: Restoration vs. Glorification

The question to the skeptic of transhumanist ideas, especially to those who are Christian, is this: Where is the line? Where does human biological improvement stray from acceptable to unacceptable? We have plenty of technologies and medical advancements today that can cure diseases and ailments through technology. Would someone with a prosthetic hip be considered a transhumanist? What about those who receive laser eye surgery? What if a cure to cancer could be found through genetic editing, which is currently being researched?[277] What if a human being could be genetically edited to become completely resistant to all diseases? Where is the line?

Many differing answers have been offered by people throughout the years, and it seems that no general consensus has been reached. Therefore, I can only speak for myself in attempting to provide an answer. My personal belief is that it comes down to the difference between restoration and glorification. Is the medical treatment restoring what was once there, or is it glorifying it into something supposedly better? Is the hip replacement an attempt to restore a person's use of his or her hip, or is it an attempt to improve the bones to the point they cannot break or wear down? Is the laser eye surgery restoring a normal, human level of vision, or is it providing someone with an ability that is unnatural to humans, such as night vision or infrared vision? Is it restoring the person to normal human ability or is it glorifying the person to superhuman ability?

That is where the line falls for me. It is found at the point where a person can no longer be considered human because he or she has advanced to something else, such as a *Homo superior*. Going further, as a Christian, I would have personal trouble accepting anything coming close to changing myself from human into something else. I would regard that as giving up my promised inheritance of glorification from God in the life to come for an inferior human and technological glorification in this life.

We will all have to stand before God someday. Even if immortality were possible through technological means, it does not change this fact: Death will still occur. Eventually, our sun will burn out. Our planet will become uninhabitable. Even if we find another planet, the universe itself will eventually come to an end. Due to the exponential expansion of the universe, a day will come when everything will be too far away from everything else to have advantageous effect. Stars will no longer be able to warm planets. Gravitational orbit will be a thing of the past. The universe itself will die. In the case that the universe ends earlier than that in a massive Higgs Field Doomsday event, even time itself will run its course. No matter what, we cannot escape entropy forever.

Transhumanism is a stall at best. No matter the level and sophistication of technology, nothing can truly live forever in this life. The next life will come for each of us. I believe it is wise to acknowledge this fact now and plan accordingly. We will be on the other side of death far longer than we could ever be on this side. I may be lucky enough to live eighty or ninety years, but after death comes, I will be on the other side until the end of time and beyond. My concern is not getting right with the technology an imperfect, flawed, human race can produce. My main concern is to be right with the Creator of humanity, the universe, and time itself, because only He can truly save me.

The good news is that He has provided a way for salvation and guaranteed glorification. It doesn't require government funding, private corporations, or waiting for advanced technology to be developed. It is available right now to every human being on earth. And even better, the God who saves desperately wants to give you this gift through His Son today, and He is only waiting for you to accept it. He wants this so much, in fact, that He gave His life for it. It cost you nothing; the price has already been paid. To truly prepare for the future, ask Jesus Christ into your life today and begin believing in and following Him.[278] Do this, and Jesus guarantees what transhumanism can never offer: You will truly and eternally be saved.

> For God so loved the world, that he gave his only begotten Son, that whosoever believeth in him should not perish, but have everlasting life. (John 3:16)

AI: Image of the Beast

By Milieu Members Paul McGuire
and Troy Anderson

He was granted power to give breath to the image of the beast, that the image of the beast should both speak and cause as many as would not worship the image of the beast to be killed. (Revelation 13:5)

What if the Antichrist isn't a charismatic world leader who seduces humanity with his spellbinding, oratorical powers?

What if the Antichrist—whom the Apostle John described as the "image of the beast"—is a godlike artificial intelligence entity, or even a transhumanist combination of a genetically altered human enhanced with artificial intelligence, a "neural lace" connected to the cloud, and possessed by Lucifer himself?

After all, Jesus Christ said in Matthew 24:37, "As it was in the days of Noah, so it will be at the coming of the Son of Man."

If God indeed destroyed the world in the Flood following the corruption of humanity's DNA when fallen angels mated with women producing hybrids known as Nephilim (see Genesis 6–9), then couldn't today's transhumanism/AI movement involving genetic modification, cybernetic augmentation, digitization of consciousness, and other forms of human enhancement be a return to the "days of Noah"?

"As well as rebuilding Babel, emerging technoscience (especially genetic engineering) is seen as recreating the transgression of the Watchers {of Genesis 6 and Enoch 1}—creating hybrid beings by crossing ontological and species boundaries in contravention of divine law—a repetition of act that will lead to a repetition of sentence," wrote Jonathan O'Donnell, a teaching fellow in the Department of Religion and Philosophies at the School of Oriental and African Studies at the University of London, in his paper, "Secularizing Demons: Fundamentalist Navigations in Religion and Secularity," from which this book, *The Milieu*, was inspired.

O'Donnell's intriguing paper, the transhumanism movement, and the recent explosion in technological advances that is altering what it means to be human comes amid increasing concerns that what Google executive Ray Kurzweil calls the "Singularity"—the moment when computers achieve human-level intelligence—could spell the doom of mankind or at least its irrevocable transformation. "2029 is the consistent date I have predicted for when an A.I. will pass a valid Turing test and therefore achieve human levels of intelligence," Kurzweil predicted. "I have set the date 2045 for the 'Singularity' which is when we will multiply our effective intelligence a billion-fold by merging with the intelligence we have created."[279]

As these dates approach, Tesla's Elon Musk is warning that humanity only has a 5–10 percent chance of preventing killer robots from destroying mankind, and Silicon Valley self-driving car engineer Anthony Levandowski has created the "First Church of Artificial Intelligence," laying the foundation for a new world religion and its AI savior.

As transhumanism and artificial intelligence are capturing the world's imagination, perhaps it's time to ask whether our traditional interpretation of the Bible's end-time scenario should consider these dizzying technological advances—and what it means for our future as members of the human race.

Stranger Beasts

The exact explanation of the phrase—"image of the beast"—in Revelation 13:5 has been debated by biblical scholars for two thousand years.

The verse begins with the words, "He was granted power to give breath to the image of the beast." The word "he" is referring to the False Prophet, who is also known as the second Beast.

The first Beast is the Antichrist, whom Satan indwells and who heads up the one-world government described in the book of Revelation.

"There is no doubt that the beast of the book of Revelation is the Antichrist that the apostles Paul and John describe in their epistles," wrote *Left Behind* series coauthors Dr. Tim LaHaye and Ed Hindson in *The Popular Encyclopedia of Bible Prophecy.*

> The world *beast* (Greek, *theerion*) means "wild animals," and Revelation first mentions this being in 11:7, where he comes from the abyss and kills the two witnesses of the Lord who convict the world of its evil…. The beast comes from the nations ("the sea," 13:1) and entices the world to follow after him (verse 3). The religious False Prophet who is also called "a beast" (verses 11–12), leads the world in worshipping the first beast and even makes an image of him (verse 15). He is identified by a mysterious number: 666 (verse 18; 16:2). The exact meaning of this number has escaped the many speculations of scholars. When the events of the Tribulation take place, and the beast comes on the scene, true Christians will probably be able to recognize him by some feature that is identified by 666…. John's beast is not simply a force, nation, or power; the beast has all of the characteristics of personality. He is the other "horn," or the "little horn" who comes out of the nations, as Daniel describes, who possesses "eyes like the eyes of a man, and a mouth uttering great boasts"—meaning he is worldly-wise and controls the nations by what he says (Daniel 7:8). This is the same beast John describes

in Revelation. The Antichrist is called "the beast" approximately
32 times in the New Testament—and only in the Apocalypse....
The Apostle Paul mentions him in much detail in 2 Thessalonians 2 as the lawless one whose coming is energized "in accordance with the activity of Satan, with all power and signs and
false wonders" (2 Thessalonians 2:9).[280]

The first Beast, or Antichrist, sets himself up to be worshipped as
God by sitting on a throne in the Holy of Holies, which is in the rebuilt
Jewish Temple in Jerusalem, and demands that he is worshipped as God.

The prophet Daniel records part of this in Daniel 9, where this event
is predicted in the last days, or the "Time of Jacob's Trouble." Both the
Apostle Paul and Christ refer to this as the "abomination of desolation,"
which Daniel describes as the Antichrist sitting in the holy place during
the "Time of Jacob's Trouble."

A variety of significant matters surround this major prophetic event.
First, we must remember that Satan, or Lucifer, is actively leading a
rebellion against God with one-third of the angels, or fallen angels, along
with all the people on planet earth who are following him, including all
those who will ultimately choose to receive the mark of the Beast (possibly a microchip implant, biochip, nanochip or DNA-modified chip).

The False Prophet is responsible for administrating the false one-
world religion and economic system that brings all this together. During
the Antichrist and False Prophet's reign of terror, no one will be able to
"buy or sell," or participate in the global economic system, unless they
have publicly renounced Jesus as Lord and have professed worshipping
the Antichrist as God.

It's only after this profession that they will be able to receive the mark
implant in their right hand or forehead, which will likely be integrated
into their DNA, allowing them to "buy and sell."

Many theologians believe the Antichrist will publicly come to power
during the beginning of the seven-year Tribulation period.

Prophecy scholars who believe in a pre-Tribulation Rapture argue that all true Christians will be supernaturally removed from the earth before the Antichrist is revealed and the Tribulation begins.

When the Antichrist rises to power and the False Prophet begins the one-world economic system and religion, anyone who refuses to renounce Christ as Lord and publicly pledge to worship the Antichrist as God will not be allowed to participate in the global economic system.

Sadly, the Bible tells us that the penalty for refusing the mark of the Beast will be having your head chopped off. Those Bible scholars who believe in a pre-Tribulation Rapture are convinced that when God promises His people in 1 Thessalonians 1:10 to "deliver us from the wrath to come" that they will be removed, or "snatched away," from planet earth before the Antichrist is revealed and the mark of the Beast system goes into effect.

A Regime Change in Heaven?

Satan has been plotting a regime change in heaven ever since he became the temporary god of this world after tempting Adam and Eve to disobey God in the Garden of Eden. That's when the Fall of Man began.

Satan's plan from the beginning has been to remove God and set himself up in the throne room of God and be worshipped as God. Satan plans to raise up the False Prophet, who very well may be alive today, to initiate a supernatural and technological plan of deception that will powerfully deceive the world's population into believing the Antichrist is God.

> And he deceives those who dwell on the earth by those signs which he was granted to do in the sight of the beast, telling those who dwell on the earth to make an image to the beast who was wounded by the sword and lived. He was granted power to give breath to the image of the beast, that the image of the beast should both speak and cause as many as would not worship the image of the beast to be killed. (Revelation 13:14–15)

Satan will give the False Prophet the supernatural power of personal and mass deception by enabling him to perform great supernatural signs, including calling down fire from heaven in the sight of all men. In addition, the performance of these supernatural signs and wonders will powerfully deceive people on earth, including even "the very elect" (Matthew 24:24).

It's important to understand that these signs may be a combination of supernatural power and advanced technologies. For example, it's been claimed that the government has holographic technologies that could cause people to witness some type of UFO invasion coming from the skies.

People involved in secret government programs have allegedly been responsible for creating purported alien encounters and people witnessing alien spacecraft. When I (Troy) was a reporter at the *Nevada Appeal* in the early 1990s, I recall a reporter at the *Las Vegas Review-Journal* telling me that his brother worked at Area 51 and they would fly experimental aircraft around Nevada, "buzzing" residents to trick them into thinking they were seeing UFOs.

The narrative of an invasion by UFOs from outer space is one that the government and Hollywood have carefully inserted into the public consciousness since H.G. Wells' 1898 novel *The War of the Worlds* and Orson Welles' radio drama, *The War of the Worlds*, in 1938 that created a public panic that earth had been invaded by extraterrestrials.

Since then, many global leaders have talked about how an alien invasion would be the fastest way to unify planet earth and bring in a one-world government. This includes American philosopher John Dewey, father of public education, and presidents Ronald Reagan and Bill Clinton. Countless movies have conveyed this idea, including Steven Spielberg's *Close Encounters of the Third Kind.*

The government, media, and Hollywood have spent decades softening up the public for such a scenario, as evidenced by a recent survey.

Derek P. Gilbert and Josh Peck noted in their excellent book, *The Day the Earth Stands Still,* that a National Geographic Channel study in 2012 found that 36 percent of Americans believe that UFOs exist, and the same percentage says they believe aliens have visited earth. In comparison, while 73 percent of American adults describe themselves as "Christian," only 10 percent now have a biblical worldview, a Barna Group survey found. "In other words, doctrinally-sound, Bible-believing Christians are outnumbered in America by ET believers three to one," Gilbert and Peck wrote.[281]

Projects Blue Beam and Montauk

One of the purported government programs linked to this belief in recent decades, and depicted on many television programs, is known as Majestic 12. The Federal Bureau of Investigation describes Majestic 12 as "a secret committee created to exploit a recovery of an extra-terrestrial aircraft and cover-up this work from public examination." However, the FBI notes that a US Air Force investigation determined that the famous document purporting to provide evidence for the existence of Majestic 12 is a "fake."[282]

Nevertheless, while the FBI discounts the existence of Majestic 12, *New York Times* bestselling author and investigative journalist Annie Jacobsen's book *Operation Paperclip: The Secret Intelligence Program That Brought Nazi Scientists to America* offers ample evidence about a secret government program that brought thousands of Nazi rocket, mind-control and genetic scientists to America following World War II. Interestingly, one of the most famous of these scientists, Wernher von Braun, became director of NASA's Marshall Space Flight Center and played a prominent role in America's space exploration program, which has contributed to growing interest in whether intelligent life may exist on other planets.

Following a recent *New York Times* story revealing the existence of a $22 million program in the Pentagon that investigates UFOs, speculation grew on Twitter, Reddit, and other online forums about the possibility of a fake alien invasion to manipulate the public. "It's tough to comprehensively explain the false flag myths without tumbling down rabbit holes, but most of them seem to stem from the writing of a prolific conspiracy theorist named Serge Monast. According to Wikipedia—the most reputable source I managed to find on the subject—Monast was a self-styled journalist from Quebec who wrote an influential paper in 1994 called 'Project Blue Beam,' which alleged that NASA and the United Nations are working together to simulate an invasion from space using holograms," Aaron Mak wrote in a *Slate* article. "The world's populace, apparently, will then abandon all religions and take this hologram to be a messiah, establishing a new worldwide religion that will serve as a string for nefarious puppet masters in the government to set up a global dictatorship called the New World Order."[283]

Researchers have speculated that Project Blue Beam may involve technologies that could create a virtual reality scenario featuring a fake rapture or the Second Coming of Christ, or the appearance of Buddha or Muhammad in the skies in which millions of people would be convinced that it's real.

Another intriguing government program that has captured public interest in recent years is the Montauk Project. This top-secret program allegedly involves mind-control and time-travel experiments. The Netflix series, *Stranger Things*, was originally titled "Project Montauk" after the covert government operation on Long Island that allegedly used runaway children as test subjects. Nazi scientists are rumored to have conducted mind-control and time-travel experiments at the facility to develop various weaponry and technologies.[284]

If these technologies exist, Operation Blue Beam could theoretically generate giant holographic projections of anything from the Antichrist becoming God ("image of the beast") to engineering a collective mass shift in consciousness where the people of planet earth believe they are becoming one and evolving into gods.

Many prophecy scholars believe the New World Order may attempt to use such technologies to overcome any resistance to or rejection of their global government. Some researchers claim this may involve scalar waves and electromagnetic radiation and wireless technologies that can purportedly transmit images and voices directly into the human brain. These investigators believe technologies could create various Christian and New Age images and sounds that could be heard by people throughout the world. These technologies could also be used by the Antichrist to communicate in direct and powerful ways to people.

"We know Jesus warned us that in the end times 'fearful sights and great signs shall there be from heaven' (Luke 21:11), and that these would be so dramatic that they will cause 'upon the earth distress of nations, with perplexity…. Men's hearts failing them for fear, and looking after those things which are coming on the earth; for the powers of heaven shall be shaken' (25, 26). It will be a terrifying time unlike any other in human history, with widespread panic and evil," Dr. Kevin Clarkson, host of the *Prophecy in the News* television show, wrote in an article in the ministry's magazine. "It is not unreasonable to believe that 'unidentified' objects from the skies could be a key part of the great end-times delusion that will engulf the entire world and lead people into following Satan in a one-world belief system and religion. This will be demonic deception at its very worst, promising a false deliverance that is going to lead to final damnation. The effects of peer pressure, groupthink, and masterful mass psychology will be manipulated by Satan's leader to ensnare the masses."[285]

A Deadly Wound

The Antichrist is depicted as being killed in the book of Revelation from a deadly head wound. Yet, somehow, he experiences a counterfeit resurrection.

When the book of Revelation states that mankind worships the "image of the beast," what kind of image is the Apostle John talking about? Does the phrase refer to a lifelike holographic projection as in

Project Blue Beam, virtual reality, the consciousness of the Antichrist uploaded into a pre-cloned body or a life-like robot, android or cyborg, or other kinds of transhumanist technologies?

> Then I saw a second beast, coming out of the earth. It had two horns like a lamb, but it spoke like a dragon. It exercised all the authority of the first beast on its behalf, and made the earth and its inhabitants worship the first beast, whose fatal wound had been healed. And it performed great signs, even causing fire to come down from heaven to the earth in full view of the people. Because of the signs it was given power to perform on behalf of the first beast, it deceived the inhabitants of the earth. (Revelation 13:11–13, NIV)

In these verses, we read that the people will worship "the image of the beast" along with the "second beast"—the False Prophet. In Revelation, we learn that the False Prophet has the power to perform great signs, including calling down fire from heaven in the sight of the world.

Does this imply interaction with UFOs, aliens, demons, or Project Blue Beam technologies that create holographic, virtual reality, or super-real displays of spaceships, aliens, the Second Coming and the Rapture in a demon-inspired, man-created narrative designed to deceive the masses into accepting the rule of the Antichrist and a one-world government?

Mark of the Beast—666

In Revelation 13:16-18, we read that the second Beast, or False Prophet, rules over a global economic system and religion.

> He causes all, both small and great, rich and poor, free and slave, to receive a mark on their right hand or on their foreheads, and

that no one may buy or sell except one who has the mark or the name of the beast, or the number of his name. Here is wisdom. Let him who has understanding calculate the number of the beast, for it is the number of a man: His number is 666. (Revelation 13:16–18 NKJV)

Just as money is spiritual today, it will be even more so during the Tribulation. If you look at the back of a US $1 bill, you'll see all kinds of occult and satanic symbols such as the all-seeing eye of Horus.

The pyramid is an occult symbol and the words in Latin below the pyramid read, *Novus Ordo Seclorum,* or "New Order of the Ages," which many researchers interpret as "New World Order." These researchers believe the bird on the left-hand side of the dollar is not an eagle, but a phoenix. This represents the legendary phoenix that is burned to death, but rises, or resurrects, to life from the ashes.

Sir Francis Bacon, the lord chancellor of England, head of the Rosicrucian Order, and the author of the *New Atlantis* who played an instrumental role in America's creation in the early 1600s, planned for America to be head of the New World Order, or the "New Atlantis." The back of the U.S. dollar is embedded with numerology and many occult symbols because it is an extension of the monetary system developed in what Revelation 17–18 calls "Mystery, Babylon." As we revealed in our bestselling globalism expose *The Babylon Code* and the second book in the series tracking unfolding end-time events, *Trumpocalypse,* "Mystery, Babylon" originated in ancient Babylon under Nimrod, ruler of the first one-world government, religion and economic system.

In Revelation 17, we read about the return of "Mystery, Babylon" and the global government, religion, and economic system that comes with it that began at the Tower of Babel.

Then one of the seven angels who had the seven bowls came and talked with me, saying to me, "Come, I will show you the judgment of the great harlot who sits on many waters, with whom the kings of the earth committed fornication, and the inhabitants of the earth were made drunk with the wine of her fornication." So, he carried me away in the Spirit into the wilderness. And I saw a woman sitting on a scarlet beast which was full of names of blasphemy, having seven heads and ten horns. The woman was arrayed in purple and scarlet, and adorned with gold and precious stones and pearls, having in her hand a golden cup full of abominations and the filthiness of her fornication. And on her forehead a name was written: MYSTERY, BABYLON THE GREAT, THE MOTHER OF HARLOTS AND OF THE ABOMINATIONS OF THE EARTH (Revelation 17:1–5, NKJV)

During this time, the False Prophet will usher in a worldwide economic system that is directly connected to the global religion. Every person on planet earth will be required to receive the mark of the Beast if he or she wants to buy or sell.

To participate in the economic system, receive health care and other benefits, you will have to receive the mark of the Beast. However, if you choose to accept the mark of the Beast and worship the Antichrist, Revelation 14 describes an angel in heaven warning mankind of their fate.

Then a third angel followed them, saying with a loud voice, "If anyone worships the beast and his image, and receives his mark on his forehead or on his hand, he himself shall also drink of the wine of the wrath of God, which is poured out full strength into the cup of his indignation. He shall be tormented with fire and brimstone in the presence of the holy angels and in the presence of the Lamb. And the smoke of their torment ascends forever and ever; and they have no rest day or night, who worship the beast and his image, and whoever receives the mark of his name." (Revelation 14:9–11, NKJV)

Elon Musk and the "AI Apocalypse"

Recently, Tesla CEO Elon Musk, speaking to US governors, warned that artificial intelligence poses an "existential threat" to human civilization.[286]

Musk used the example of computer game technology, such as virtual reality, that is already approaching a point that it is indistinguishable from reality. "If you assume any rate of improvement at all, then the games will become indistinguishable from reality, just indistinguishable," he said. "Even if the speed of those advancements dropped by 1000, we are clearly on a trajectory to have games indistinguishable from reality."

Musk went on, saying that people should be very concerned about AI. "I keep sounding the alarm," Musk says. In 2014, Musk compared "A.I. developers to people summoning demons they think they can control." The following year, he signed a letter warning of "the risk of an A.I. arms race." In early 2017, Maureen Dowd wrote an article in *Vanity Fair* about Musk's "crusade to stop the A.I. Apocalypse."

"In a startling public reproach to his friends and fellow techies, Musk warned that they could be creating the means of their own destruction," Dowd wrote. "He told Bloomberg's Ashlee Vance, the author of the biography *Elon Musk*, that he was afraid that his friend Larry Page, a co-founder of Google and now the C.E.O. of its parent company, Alphabet, could have perfectly good intentions but still 'produce something evil by accident'—including, possibly, 'a fleet of artificial intelligence-enhanced robots capable of destroying mankind.'"[287]

AI Salvation and "Way of the Future"

Society's technological advances, such as artificial intelligence, are accelerating into hyperdrive, and transhumanists like Musk and others are expressing growing concerns that when artificial intelligence surpasses human intelligence that computers, robots, androids and cyborgs could potentially rule over mankind.

Ironically, the humanistic society that produced transhumanism is based on Charles Darwin's theory of evolution in which the fittest survive. All evolutionary theory is based on the belief that genetically stronger and more intelligent species will rule over or make extinct inferior species.

This is ironic because the pervasive fear among many transhumanists is that man's creation of AI is self-evolving and will become stronger and more intelligent than mankind. As we've noted, Kurzweil predicts that the Singularity will occur around 2045 when AI computers, robots, and androids will become superior to mankind. Musk has warned that when machines can outthink humanity, this could be the end of humanity.

In what seems like an effort to stay on the good side of their new god, Levandowski has created what *Wired* called the "First Church of Artificial Intelligence." "The new religion of artificial intelligence is called Way of the Future," Mark Harris wrote in *Wired*. "Papers filed with the Internal Revenue Service in May (2017) name Levandowski as the leader (or 'Dean') of the new religion, as well as CEO of the nonprofit corporation formed to run it. The documents state that WOTF's activities will focus on 'the realization, acceptance, and worship of a Godhead based on Artificial Intelligence (A.I.) developed through computer hardware and software.' That includes funding research to help create the divine A.I. itself."[288]

Image of God and Soulless AI

Humanists and transhumanists reject the biblical account of Creation that says God created man in His own image. In Genesis 1, we read how God created the first man and woman:

> So, God created man in His own image; in the image of God
> He created him; male and female He created them. Then God

blessed them, and God said to them, "Be fruitful and multiply; fill the earth and subdue it; have dominion over the fish of the sea, over the birds of the air, and over every living thing that moves on the earth." (Genesis 1:27–31, NKJV)

What we can learn from this is that men and women have the DNA of God. Originally, in the Garden of Eden, Adam and Eve had the completely pure and uncorrupted DNA of God. Then when they were tempted by Lucifer they disobeyed God's only commandment and released the death force—activating the law of sin and death into all of creation and humanity resulting in the Fall of Man.

Humanists scoff at the Fall of Man as a pathetic fairy tale. But ever since the Fall, Adam and Eve experienced the degradation of their DNA that allowed the death force to change them and the world.

However, even in man's fallen state, he is still capable of God-like achievements in terms of creativity, science, and technology. In addition, every man and woman have a soul or spirit completely distinct from their biological bodies. Just because science has not sufficiently evolved to develop technologies to quantify the soul—measure it, photograph it, analyze its bioenergetics field and the frequencies that compose it and generate it—doesn't mean it doesn't exist.

Another factor is that all men and women alive have the law of God written on their hearts, whether they choose to ignore or suppress it. The law of God written on their hearts, or inner being, is the programming of God that runs the software of God, which is the body, soul, and spirit.

Conversely, transhumanism is creating self-evolving machines that are soulless through artificial intelligence. In a very real sense, AI computers, robots, and androids are made in the image of man. But AI made in the image of man operates from reason, logic, pragmatism, and so-called scientific empiricism. This means evolving AI machines will have no conscience, no sense of right or wrong or genuine compassion, and they will not be able to know and express the agape, selfless love of Jesus Christ. AI machines will always be soulless, because only God can give

His creation, mankind, a soul. All other species of life on planet earth are soulless.

The truth is that a very real being, Lucifer, has wanted to be God since the beginning. Lucifer's primary temptation to mankind is, "ye shall be as gods" (Genesis 3:5). Lucifer wants to rule over earth and be worshipped as God. The great biblical mystery the Apostle John described in Revelation 17–18—Mystery, Babylon—reveals how Satan plans to accomplish this.

Mystery, Babylon is simply an occult religion based on Luciferian programming. Mystery, Babylon birthed the world's first one-world government, religion, and economic system, and will return in the last days as the New World Order—the emerging global government, cashless society, and universal religion.

All this is driven by Lucifer and his followers. It's a mechanism for allowing Lucifer to replace God, set himself in the rebuilt Jewish Temple in Jerusalem, and demand that all people on earth worship him. In addition, through the False Prophet, a global economic system will be created that requires people to take the mark of the Beast.

Although transhumanists are operating by pure faith since they have no scientific data to predict precisely where AI and self-evolving machines are going, it's critical when attempting to analyze the future to understand that planet earth, all of creation, mankind, and any technologies that mankind creates, are under the invisible restraints of the personal living God of the universe and His laws.

It may temporarily and superficially appear as if transhumanists and the scientific creators of AI have unlimited freedom to create whatever they can, but the biblical God is still the Supreme Being.

Return of Days of Noah

An example of this is the account of Noah's Flood. A growing number of Bible scholars believe Genesis 6 tells us that, prior to the deluge, fallen angels mated with human women, producing a hybrid species known as Nephilim, in which the DNA of fallen angels was mixed with the DNA of human men and women.

The fallen angels needed physical bodies to inhabit the physical realm of the earth, and they needed bodies to experience the pleasures of human bodies. As such, fallen angels who possessed an advanced knowledge of science, technology, mathematics, and other fields were the beings who helped man build ancient super-civilizations that many researchers now believe existed in the antediluvian world.

In their experimentations, they reproduced with various species of animals. The ancient legends of mermaids, elves, half-human and half-horse beings, and fish-like creatures that were both men and fish are based on this.

When we study Plato's account of Atlantis, we see that he believed the philosopher kings who ruled Atlantis came from ancient beings that were part fish and part human. Whether Plato's account is just a legend is open to debate. The point is that the fallen angels had corrupted the DNA of every species on earth and all of mankind except for Noah, his sons, and their wives.

God specifically instructed Noah to build an ark that would hold two of every kind of animal, both males and females. The Flood of Noah wiped out every living thing except for the animals, Noah, his wife, his sons, and his sons' wives. After the Flood wiped out the corrupted DNA, God commanded Noah to "be fruitful, and multiply" (Genesis 1:28).

Why did this global catastrophic event happen? It happened because humans crossed over some invisible line wherein women began mating with fallen angels. Throughout the Bible, we read about bloodlines, DNA, genetics, etc. When man tampered with the DNA of God, the Flood came.

Then the LORD said to Noah, "Come into the ark, you and all your household, because I have seen that you are righteous before me in this generation. You shall take with you seven each of every clean animal, a male and his female; two each of animals that are unclean, a male and his female; also seven each of birds of the air, male and female, to keep the species alive on the face of all the earth. For after seven more days I will cause it to rain on the earth forty days and forty nights, and I will destroy from the face of the earth all living things that I have made." So, Noah, with his sons, his wife, and his sons' wives, went into the ark because of the waters of the flood. Of clean animals, of animals that are unclean, of birds, and of everything that creeps on the earth, two by two they went into the ark to Noah, male and female, as God had commanded Noah. And it came to pass after seven days that the waters of the flood were on the earth. In the six hundredth year of Noah's life, in the second month, the seventeenth day of the month, on that day all the fountains of the great deep were broken up, and the windows of heaven were opened. And the rain was on the earth forty days and forty nights. On the very same day Noah and Noah's sons, Shem, Ham, and Japheth, and Noah's wife and the three wives of his sons with them, entered the ark—they and every beast after its kind, all cattle after their kind, every creeping thing that creeps on the earth after its kind, and every bird after its kind, every bird of every sort. And they went into the ark to Noah, two by two, of all flesh in which is the breath of life. So those that entered, male and female of all flesh, went in as God had commanded him; and the LORD shut him in. Now the flood was on the earth forty days. The waters increased and lifted up the ark, and it rose high above the earth. The waters prevailed and greatly increased on the earth, and the ark moved about on the surface of the waters. And the waters prevailed exceedingly on the earth, and all the high hills under the whole heaven were covered. The waters prevailed fifteen cubits upward, and the mountains were covered. And all flesh died that moved on the earth: birds and cattle and beasts and every creeping thing that creeps on the earth, and every man. All in whose nostrils was the breath of the spirit of life, all that was on the dry land, died. So, he destroyed all living things which were on the face of the ground: both man and cattle, creeping thing and bird of the air. They were destroyed from the earth. Only Noah and those who were with him in the ark remained alive. And the waters prevailed on the earth one hundred and fifty days. (Genesis 7:1–24, NKJV)

Notice in the above passage verse 22: "All in whose nostrils was the breath of the spirit of life." God put the "breath of the spirit of life" into Adam and Eve, all mankind, and all life. The "breath of the spirit of life" is the divine energy of the Creator's life force, and it is what separates what God has created from AI computers, robots, androids, and cyborgs, which, despite their amazing capacity to think and reason, don't have the life force and "the breath of the spirit of life" in them.

These AI machines will remain soulless, which should terrify their creators and all mankind, because they will become lifelike beings just like humans and possess many powers, but they are soulless.

God's Word teaches us that Satan directed two hundred of his fallen angels to descend upon Mount Hermon to mate with human women and give mankind various occult powers and advanced sciences and technologies.

Below Mount Hermon was the ancient civilization of Phoenicia, whose seagoing traders spread this knowledge across the world. The connection between Mount Hermon and Phoenicia explains why the New World Order is closely associated with the mythical phoenix, which some researchers believe is the symbol on the back of the US dollar, not an eagle.

New Atlantis, "Gate of God," and CERN

When Bacon planned for America to be head of the New World Order and the "New Atlantis," he was aware of these themes. The problem is that the New World Order will be headed by the Antichrist, and the "New Atlantis," according to Plato, is the model of a society ruled by a scientific elite with a genetic heritage going back to Mount Hermon and Phoenicia.

But the overwhelming area of concern is that just like the hybrid species of fallen angels and human women produced a soulless race, that soulless race possessed advanced technologies and initiated the satanic and pagan worship of Baal, Ashtoreth, and other gods that dominated the land of Canaan under the Nephilim giants.

Soulless beings, by their very programming, automatically come under the rule of Lucifer, because Lucifer, his fallen angels, and mankind who choose to serve him, are either soulless, or the life force of God, which is the Holy Spirit, is deactivated in humans who consciously reject God's free offer of salvation in Christ.

A Dimension outside Time and Space

When men or women are not born again, their souls are completely dead, and they are dead to the things of God.

When by faith they receive Christ into their lives and are born again, the Holy Spirit enters their being and supernaturally regenerates their inner being, and they become brand new creatures in Jesus.

The spirit of God regenerates their soul, and the life force of God causes them to become alive. The result of this is that although their physical bodies are dying, they now have eternal life. And, when they die, they will go immediately into the presence of the Lord in a completely different dimension outside of time and space called heaven.

They will have no need of AI-evolved machine bodies like robots, androids, and cyborgs, in which their consciousness can be uploaded, and they can live in the true virtual reality of eternal life in heaven.[289]

The Eschatological Awareness of Popular Science Fiction

By Milieu Member Frederick Meekins, PhD

In the article, "Secularizing Demons: Fundamentalist Navigations in Religion and Secularity," S. Jonathon O'Donnell of the University of London conveys the impression that the primary hindrance to the ultimate triumph of the utopian transhumanist agenda is a small clique that he has given the nefarious title of "The Milieu," composed of what he categorizes as "a loose network of post-denominational evangelical bio-conservatives (642)." From that, the average reader is likely to assume that the sorts of concerns raised by such Christian scholars regarding subjects considered somewhat fringe in nature, such as human enhancement, life beyond this world, sentience technological in origin, and the implications of such in the areas of reflection considered religious in character have been concocted solely by those suffering from a shared but rare form of philosophical psychosis. Though it could be argued that those of this post-denominational mindset spanning a variety of theological interpretative commitments ranging from the charismatic to the conservative evangelical to the fundamentalist are perhaps the most deliberately systematized in articulating their opposition in terms of worldview to these sorts of revolutionary technologies, such sentiments do not necessarily find their most widespread cautions voiced by what must be admitted is a rather isolated handful of religionists inclined towards speculative eschatological prognostication. Rather, these sorts of themes are just as likely to be found in the works of popular science fiction surprisingly even echoing the warnings raised by the more explicitly theological works.

Broadly speaking, science fiction consists of works of the imagination considering the implications of advances in technology or ideas capable of altering the very paradigms or underlying assumptions through which human beings perceive reality. As such, often these narratives focus as much (or even more so) upon the resultant sociology rather than the mechanics of the technology bringing about such comprehensive revolutionary change. Also just as worthy of note is that the intention of these "tales of suspense" (to borrow the words of an old *Marvel Comics* title that featured the exploits of Captain America, Iron Man, and—interestingly enough—a cosmic being named The Watcher, whose stated mission was to observe but never interfere, despite his doing so at various critical moments in human history) has not always been simply to contemplate in a manner of dispassionate detachment akin to a sort of reverse historian (for lack of a better term) how the world of tomorrow might be organized and function. Often the purpose of such creative undertakings has been either to comment on something in the present that has caught the attention of a particular author or even to advocate on behalf of a particular perspective in the hopes of advancing it.

The conventions of the genre as now known were formulated roughly from the mid-1800s through the early decades of the twentieth century in works published by esteemed figures of literary history such as Jules Verne and H. G. Wells and further elaborated upon by writers who would be considered more contemporaneous, such as Isaac Aşimov and Robert Heinlien. On the surface, science fiction of the period seemed to celebrate the modernist assumption that there was nothing rationalistic empirical science could not ultimately understand or explain in terms of materialistic phenomena. That which could not be was to be tossed aside as of little consequence or even altered through new knowledge gained as a result of breakthroughs in fields such as pharmacology. Arthur C. Clarke, author of classics such as *2001: A Space Odyssey*, quipped along the lines that what is understood as magic is merely technology so advanced as not yet to be understood.

Despite the pervasiveness of secularity pushing against the boundaries of traditional beliefs to the point of even causing their erosion, a religious impulse of some kind remains one of the most significant influences even if the contents of such have been altered beyond recognition. Man simply cannot disentangle himself from the inclination to perceive reality through this particular lens of perception where the purpose of the species is to be derived from a source beyond himself. Even if prominent luminaries in the field of science fiction professed explicit hostility towards traditional formulations of orthodoxy (Issac Asimov to the point of serving as the president of the American Humanist Association), the narratives promulgated by the voices inclined towards this form of literary speculation could not escape how the high technology uplifted as the cure all for the struggles plaguing humanity would attempt to fill this void they themselves inflicted in part upon the human heart, even if these substitutions would prove to be disastrous even within the context of the stories themselves.

One of the deepest longings found in some form or another in nearly all religions, at least to some degree, is for a prophesied deliverer of extraordinary ability to arrive from on high, often through origins beyond the ordinary, or as a result of a combination of the two. In Hinduism, such a figure is known as an avatar, often of the god Vishnu such as Krishna. In Buddhism, a corresponding figure with the purposes of bringing enlightenment into the world is referred to as a Bodhistatva. In Islam, certain sects await the arrival of either the Madhi or the Last Imam to usher in that particular form of theism's version of the end times.

This idea of the anointed one no doubt found its most complete expression in the form of the Messiah still anticipated by devout Jews but believed by Christians to have already completed His initial and most metaphysically profound work upon the earth through the death, burial, and resurrection of the Lord Jesus Christ, who promised to soon return to culminate all of history. As to the listed criteria, Jesus could be said to have arrived from on high as Christians holding to expressions of the faith considered orthodox believe Him to be a member of the triune Godhead along with God the Father and God the Holy Spirit. The doctrine of the Trinity finds justification in a number of biblical passages. In John 10:30, Jesus says, "I and my Father are one." The Holy Spirit is believed to be a member of this most perplexing of ontological enigmas on the basis of Matthew 28:19, from which the baptismal formula of "in the name of the Father, and of the Son, and of the Holy Ghost" is derived. The origin or conception of Jesus could be said to be beyond the ordinary in that Isiah 7:14 foretold that a virgin would conceive, with this promise fulfilled in Luke 1:26–39, when the angel Gabriel appeared before Mary to announce that she had been selected to fulfill this holy purpose.

Likewise, the adherents of the so-called great religions are not the only ones insisting that circumstances are so dire that a messianic figure must intervene to restore the world and return it to some degree of normalcy. This is also a theme quite common throughout the annals of speculative literature such as science fiction. Some, interestingly, even share parallels with a number of biblical accounts warning of the destruction that will result when faith is placed in these figures when they are really not who they claim to be.

It could be argued that the era of costumed adventurers possessing powers and abilities far beyond those of mortal men (to borrow the phraseology of the classic television series featuring the character) began with the debut of Superman in the pages of *Action Comics* in June 1938. Hailing from the planet Krypton, this character rocketed to earth, sent by his father following the destruction of his home world. In many versions of the story retold numerous times over the decades, the infant in the spaceship is found after the vehicle crashes in a Kansas field by salt-of-the-earth, all-American couple Jonathan and Martha Kent just moments after offering up a prayer for a child upon learning that they were unable to have a biological one of their own. Though appearing superficially human, because of his extraterrestrial physiology, Superman is able to absorb energy from the earth's yellow sun to fuel a number of powers such as super strength, heat vision, imperviousness to bullets, and flight (probably the aptitude that most captures the imagination of young and old alike).

Though many Superman stories deal with typical comic book escapades, such as battling a number of bad guys (many also similarly super-powered), trying to disguise himself as Clark Kent so that he might enjoy some semblance of a normal life, and the resulting comedy in which the girl of his dreams is for the longest time in love with him as Superman but does not have the time of day beyond their workplace relationship at the *Daily Planet* as Clark Kent, the saga continues to resonate with audiences around the world when a number of characters nearly as popular from that time, such as the Phantom or the Shadow, have for the most part been forgotten by all but the most hardcore fans. It can be assumed that something about the so-called Man of Tomorrow must therefore touch upon a number of themes or symbols deeply ingrained upon the human soul.

In *The Gospel According to the World's Greatest Superhero*, Stephen Skelton makes note of a number of striking parallels between the Son of God and the Man of Steel. For example, the Kryptonian family into which Superman was born is named "El." Skelton writes, "Superman and his father share the last name of *El*—the Hebrew word for *God*. Thus in the Superman story, when 'El' the father sends 'El' the son down to Earth, 'God' the father sends 'God' the son down to Earth (20)." If that is not enough to at least prick the ears of the discerning, Skelton further points out that Jonathan and Martha Kent were also originally intended to instead be named Mary and Joseph until publishers thought the better of it.

The parallels between these two figures does not stop there. Taking the Hebrew linguistics to their ultimate conclusion, according to a June 4, 2013, *Times of Israel* story titled, "Man of Steel No Longer 'Man of Shtell,'" Superman's Kryptonian name Kal-El can actually be translated as "the Voice of God." Of Christ, John 1:1 says, "In the beginning was the Word, and the Word was with God, and the Word was God."

Without a doubt, the creators of Superman most likely drew upon these cultural sources in part for inspiration. Jerry Siegel and Joe Schuster were two Jewish youths from Cleveland with an interest in the science-fiction pulp magazines of the day. Jerry Siegel just a few years earlier had lost his father in an armed robbery of the family store, which no doubt accounted for the character being bulletproof. He would later reflect that Superman was in part inspired directly by the Old Testament account of Sampson (Skelton, 37). But in creating a champion of the downtrodden and oppressed fighting the proverbial never-ending struggle for truth, justice, and the American way, the character could not help but be construed as a sort of secular messiah.

Even if the implications of such were glossed over in the early years of the character in favor of drama and theatrics, these metaphysical implications made their way into the foreground as these works of spectacular imagination grew to become more existentially reflective. This trend was probably at its most deliberate in the 1978 film, *Superman: The Movie,* directed by Richard Donner with the screenplay written by Mario Puzo and Superman played by Christopher Reeve in what many fans consider the definitive portrayal. This tendency to view Superman as a bit more than just a costumed adventurer with the ability to fly is epitomized by two quotes attributed to Superman's biological father, Jor-El, portrayed by Marlon Brando.

Jor-El says in the first quote, "You will travel far, my little Kal-El. But we will never leave you...all the days of your life. You will make my strength your own, and see my life through your own eyes, as your life will be seen through mine. The son becomes the father, and the father, the son." In that, the Christian cannot help but have biblical phrases such as "I will be with you always even unto the ends of the earth" and "my father and I are one" come to mind. The second quote can be construed as something of a Messianic admonition on the part of Jor-El to his son. In it, the Kryptonian statesman counsels, "Live as one of them, Kal-El, to discover where your strength and your power are needed.... They can be a great people, Kal-El; they wish to be. They only lack the light to show the way. For this reason above all, their capacity for good, I have sent them you…my only son."

However, the problem with a secular messiah is that the figure is ultimately a naturalistic or de-supernaturalized messiah. All that a figure such as Superman can do is provide an example. He himself has proven to be as flawed as the rest of us, as those shocked by the scene in the sequel *Superman II* of this shining beacon to humanity snuggling with Lois Lane undressed under the reflective aluminum foil sheets in the Fortress of Solitude without sanction of matrimony can attest.

Within science fiction, there is indeed an influential strain that salvation will not so much consist of a forgiveness of sin and the eventual liberation from death and decay through that particular beatific route, but rather through technological enlightenment. This form of expanded consciousness is often bestowed by intelligences from beyond this world, or at least by formidable elites originating here on earth. However, just as often found within these stores are the "hermeneutics of suspicion" S. Jonathan O'Donnell attempts to pin the blame solely on Christians for intending to derail utopia at all costs.

Star Trek II: The Wrath of Khan is probably considered by both critics and Trekkers the greatest of the Star Trek films. Portrayed by Ricardo Montelbán, the character proved to be one of the most formidable adversaries faced by the crew of the *Starship Enterprise* from the original series. Seeking revenge for the loss of his wife after having been exiled by Captain Kirk on an isolated planet approximately twenty years earlier, Khan is not stopped until Spock willingly sacrifices his own life (only to have it restored in the sequel, thanks to the Genesis Device).

Yet it is the back story of Khan Noonien Singh that is even more captivating. Introduced in the episode of the original *Star Trek* series titled "Space Seed," the crew finds Khan and his associates in cryonic hibernation adrift in deep space aboard what, to Captain Kirk and his colleagues, is a very old vessel named the *Botany Bay*. It turns out that Khan and company are genetically engineered beings of augmented ability and intellect who took control of significant portions of the world during a time referred to as the "Eugenics Wars." *Star Trek* producers in the late 1960s predicted that this conflagration resulting from tinkering with the fundamental ontology of select individuals to impose a new world order would take place in the late twentieth or early twenty-first century—in other words, at about the time of this writing. The story is fleshed out in more detail in the two-part novel by Greg Cox titled *The Eugenics Wars: The Rise and Fall of Khan Noonien Singh*. In the story, Khan and others like him were developed by a secretive organization known as the Chrysalis Project for the purposes of rising into prominent positions around the globe to seize eventual power.

It has been argued that *Star Trek* has retained a devoted following since its debut in the 1960s largely because of the positive vision of the future the franchise offers wherein humanity, in cooperation with a number of similarly minded species, has resolved the vast majority of its internal issues to embrace a higher quality of life through dependence upon advanced technology rather than upon traditional conceptions of the supernatural. However, even the visionary futurists behind *Star Trek* (and not just those pesky troglodytes of The Milieu) do not believe that a technocracy as diverse and as inclusive as the United Federation of Planets would be able to get out from under the shadow of the potential threat posed by a genetic overclasss. In the episode "Doctor Bashir, I Presume" of the series *Star Trek: Deep Space Nine*, it is revealed that the station's chief medical officer Julian Bashir did not naturally possess the amount of sheer brain power necessary to master a profession as complex as interplanetary medicine. Rather, he was born mentally deficient—but his parents decided to pursue genetic treatments that led to enhancements in violation of the law in order to correct what they perceived as disabilities curtailing their son's potential and future quality of life. The drama in the episode arises not so much from the possibility of Dr. Bashir's parents being punished for what is considered a very serious crime, but from the prospects of Julian facing the possibility of having to forfeit everything he has worked to achieve because of the stigma attached to the genetic advantage he acquired through no fault of his own.

A significant percentage of the eschatological concern articulated by the scholars and analysts of The Milieu focuses upon the ascent of just such a figure of extraordinary ability who will at first dazzle the world, much like the Last Son of Krypton, but who will eventually plunge the world into a calamity sounding eerily similar to that of the Eugenics Wars alluded to occasionally throughout the history of *Star Trek* across assorted media. Christians refer to the future tyrant as the Antichrist, the Beast, or the Son of Perdition. Along with this resultant body of research, this cadre of eschatologists has also reflected deeply upon how advanced technology straddling an increasingly thin boundary between the mystical and the scientific could be used to bring about the sorts of conditions and events described in the Bible categorized by theologians as thend times.

In terms of encountering scenarios remarkably similar to those warned of by The Milieu, one does not have to look much farther than a series of loosely interconnected movies and television series commonly referred to as the Marvel Cinematic Universe. These films focusing upon the superheroes originally appearing in the pages of *Marvel Comics* began with the first *Iron Man* movie about an eccentric billionaire inventor and the advanced suit of armor that not only allows him to perform amazing feats of flight and strength, but that also in part sustains his life by preventing shrapnel embedded in his chest from damaging his heart beyond repair. Though dazzling, the Iron Man suit, in principle, is not that much different than a variety of other technologies audiences are already familiar with. Airplanes have taken to the skies since the early twentieth century, and knights took to the fields of battle wearing what they hoped were protective suits of armor centuries ago. However, by essentially combining the two and then deploying the resultant combination with his own sense of pizzazz, at the end of the film, Tony Stark is informed by the director of the secret global intelligence agency SHIELD, Nick Fury, that he has inadvertently thrown open to public view a door to a world that most had no idea existed.

From the first two films of the *Captain America* series, the viewer learns that there is an organized conspiracy attempting to exert a deliberate, concerted effort to control the outcome of geopolitical events. In *Captain America: The First Avenger*, Hydra is initially presented as a special-projects division within the Nazi hierarchy marked by a fanaticism surpassing even that of the SS. Headed by a figure known as the Red Skull, disfigured in the same sort of experiment that transgenically endowed Captain America with a variety of enhanced abilities, the goal of Hydra—echoing and expanding upon the mission of the Ahnenerbe from actual history—is to seek out the most powerful objects of Germanic mythology and to adapt these as technologies for the war effort. Eventually, Captain America foils the Red Skull's plot, but only at considerable personal cost as the hero is plunged into a cryogenic hibernation lasting nearly seventy years.

In the second film of the series, *Captain America: The Winter Solider*, viewers find the patriotic Steve Rogers in a world where the eternal verities he constantly strives to embody (much like Superman at DC Comics) are not necessarily deemed all that essential for efficient governance and are often viewed by the power elite as a hindrance to their own utilitarian agenda. Captain Rogers, with his traditional-values approach to life, finds his professional life challenging enough when his perspective clashes with that of SHIELD Director Nick Fury's own realpolitik approach to world hot spots exacerbated by an assortment of troublemakers with skills and abilities transcending those of run-of-the-mill-terrorists. However, the situation is complicated when it is revealed that Hydra was not eliminated following the defeat of Nazi Germany at the end of the Second World War. The organization continued to exist clandestinely, surreptitiously accumulating power and influence by infiltrating the highest echelons of society whose members greet one another with a hushed whisper of "Hail Hydra" in the ear in the equivalent of a Masonic handshake. As in the case of the mythological creature after which the organization is named, should any of the members happen to fall, two more stand ready to assume their comrade's place.

Even Glenn Beck praised *Captain America: The Winter Soldier* for its insight throwing open the veil as to how global politics likely operates at the highest levels in the attempt to implement a state of total security that, by its nature, must eliminate all forms of dissent and nonconformity. However, it was when the events that transpired in *Captain America: The Winter Solider* were expanded upon in the TV series, *Marvel's Agents of SHIELD* that the narrative in a sense took into consideration the principalities and powers motivating the movement towards global governance beyond the mechanics of how such an attempt to seize power might take place. In a story arc set during the third and fourth seasons of the series, it is revealed that Hydra was not so much simply a branch of the Nazi Party, but rather a secret society that predated that particular totalitarian movement by centuries. Elites belonging to this society for generations pledged themselves to a mysterious god-figure who promised to one day return to lead his followers to glory. Devotion to this being was so absolute that members of the order were to demonstrate this through sacrifice of themselves or their children if so required through a ritualized selection process wherein the victim was sent through a portal to a planet in a distant star system where the entity awaited this sustaining oblation.

What the members of Hydra perceived as its god-figure ties in with the next conceptual level of the Marvel Cinematic Universe and the eschatological concerns raised by The Milieu. That is namely the role played by extraterrestrials as a sort of substitute deity in a reality characterized by techno-spiritualism. Hydra's "god" turned out to be nothing of the sort. Rather, the individual happened to be an Inhuman with the ability to elicit a response of euphoria and devotion bordering on worship on the part of individuals exposed to this form of bioorganic mind control. In terms of the Marvel cinematic universe, an Inhuman is an individual that has been enhanced on the genetic or cellular level to exhibit some form of paranormal ability resulting through a process referred to as terragenesis. The process originated as part of an experiment by an alien species known as the Kree in order to engineer a breed of super soldiers to defend their expanding interstellar empire and resolve their own evolutionary stagnation.

Though it is not a topic as developed in the Marvel films to the same extent as the comics serving as the inspirational foundation of the movies, the Kree are not even the first extraterrestrial species to tinker with life on earth in such a transformational manner in what serves as something of a mythology for the contemporary era. In Marvel canon, human beings were created by yet another group of extraterrestrials even more powerful than the Kree, known as the Celestials from indigenous, preexisting primates. This program of directed intelligent design resulted not only in baseline human beings, but also in two additional strains of sentient life. These were the Deviants, often characterized by disturbing deformities resulting from constant mutation, and the Eternals, the epitome of the genetic manipulation on the part of the Celestials so much so that a number were thought to be gods by ancient humans.

One of the shared assumptions of The Milieu is likely that the beings understood to be from beyond this world are likely not extraterrestrials in the popularly understood sense of a life form originating on a planet not all that different than earth in terms of being in the same spatio-temporal continuum. Rather, these thinkers hypothesize that these non-terrestrial entities originated from another plane of existence or dimension altogether. Although the Marvel cinematic universe possesses an abundant number of extraterrestrials originating from planets on this particular level of the multiverse, the body of imaginative works under this particular literary umbrella also possesses that additional layer where beings exist in realms transcendent to what the human species perceives as reality.

The narrative branch taking this possibility into consideration the most seriously consists of the *Thor* series of films. One of the foundational concepts behind these posits that the entities understood to be the gods of Norse mythology such as Odin, Thor, and Loki did indeed exist. It is just that they were not gods, but rather higher-order beings from the realm of Asgard possessing both a biology and (probably even more importantly) a technology much more advanced than that of mortal earthlings.

This is especially evident in the second film of the series, titled *Thor: The Dark World*. Though admittedly not the most memorable of the Marvel cinematic antagonists when compared to Tom Hiddelston's performance as Loki or even James Spader as the malevolent artificial intelligence Ultron, the Dark Elves led by Malekith are still an intriguing adversary straddling the murky boundaries between mysticism and advanced science. It is revealed that Malekith and his followers are from a time before the universe itself, described as "before the birth of light." The average viewer of popular cinema is probably aware that in most forms of serious fantasy, elves are depicted as being far more formidable than their counterparts who bake cookies in trees or putter around in Santa's workshop. However, the Dark Elves depicted in *Thor: The Dark World* do not even adhere to the more medievalist conventions established for their species in imaginary epics such as *The Lord of the Rings*, for these particular elves shoot laser guns and fly around in gigantic spaceships that would give Darth Vader a run for his money.

Such would make Malekith and his kin more akin to the interpretation of extraterrestrial phenomena advocated by The Milieu that these entities are actually transdimensional beings. Some might think it is a bit of a stretch to view extraterrestrials as the elves of Norse or derivative Northern European mythologies. However, that might be a more traditional way to understand these sorts of malevolent intelligences rather than through the more contemporary paradigms of little green men and flying saucers. Jacques Vallee is quoted by Gary North in *Unholy Spirits: Occultism and New Age Humanism* as saying, "Why is it, I wondered, that the 'occupants' of UFOs behave so much like the denizens of fairy tales and the elves of ancient folklore? Why is the picture we can form of their world so much closer to the medieval concept of Magonia, the magical land above the clouds, than to a description of an extraterrestrial planetary environment[?] (315)."

Adherents of nearly all faiths will attempt to make the case that they are drawn to their respective creedal professions in terms of how their lives have been improved as a result of embracing the higher truths and values taught by these respective religions. However, if pressed, most will admit that what often pushed them towards faith is that nagging concern eating away at each of us to varying degrees of what will happen to us after we die and what we must do to make that existential state coming next as positive as possible. Even if materialism has done everything within its power to blunt the fear of a God who punishes sin so that the pleasures of this life might be maximized, the latest advances in science have yet to vanquish this most relentless of foes, even if for many the date with mortality has for a while at least been delayed. As such, a number of the most creative minds have turned to imagining ways in which to reconcile their metaphysical proposition that this reality is all that exists and the unshakeable desire for continued existence even after death.

One intriguing way that technologically assisted, everlasting life might be achieved was considered on *Marvel's Agents of SHIELD*. As in the case of both the other concerns raised by The Milieu and other examples of speculative fiction, the results were not all that they were hoped to be. In the series, the technology referred to as "the Framework" consisted of an advanced computer into which the human consciousness could be uploaded. The Framework was developed by scientist Holden Radcliffe (deliberately referenced in the dialog as a transhumanist) in the attempt to save the life of his dying fiancée. In the Framework, the individual is granted the opportunity to experience their idealized life. For example, upon being forced into the Framework, SHIELD agent Phil Coulson finds that he is a middle school civics teacher (the pressures of his cloak and dagger existence long forgotten) and Agent Mack Mackenzie is reunited with at least the VR (virtual reality) facsimile of his daughter who had died in the offline world. However, this bliss does not continue for very long.

Before creating what was intended as a virtual reality afterlife that would provide him with continued access to his dying fiancée, Holden Radcliffe, along with SHIELD scientist Leopold Ftiz, engineered an android based upon her appearance known as a Life Model Decoy named "AIDA" (Artificial Intelligent Digital Assistant). AIDA becomes much more than an Amazon Echo with a pretty face and an alluring exterior after being utilized as an interactive computer interface in order to access an occult text containing the secrets of the universe known as the Darkhold so powerful that it drives insane those gazing upon its pages directly. This mobile artificial intelligence having achieved self-awareness as a result uploads "herself" into the Framework, which she proceeds to recast in her own image. By doing so, she assumes the role of the head of Hydra and attempts to eliminate the SHIELD operatives slowly starting to realize that they are no longer in the real world but rather that their minds are trapped in a virtual-reality construct. And unlike the paradise provided by an eternal just and loving God where the resurrected body will be glorified, when you die in the deceptive digital construct, you also die in the everyday world of real flesh and blood.

The question can easily and legitimately be asked: Can't a story just be a story without an ulterior motive beyond that to simply entertain? The answer would be: "It depends." Films have been produced to advocate all sorts of agendas and perspectives. However, Hollywood is, in essence, a business in pursuit of a profit over and above nearly everything else. To achieve blockbuster or iconic status, a narrative or production must touch upon a truth that a significant number of people are aware of on some level, even if they do not understand fully what they are being presented. The analysts of The Milieu attempting to comprehend unfolding events in light of prophetic truth and the literary creatives responsible for these journeys into phantasmagoria often possess worldviews in diametric opposition regarding the purpose of man and the origins of the universe. However, despite these varying interpretations, these are often in congruence in hypothesizing that the advanced technologies being developed will likely assist in unleashing the greatest of tribulations that the world has ever known.

The Vatican Imagines a Pro-Transhuman Milieu of Its Own

By Milieu Leader Dr. Thomas R. Horn

I n November 2017, the Vatican's Pontifical Council for Culture hosted a plenary assembly on "The Future of Humanity—New Challenges to Anthropology" that included top-level scientists and cardinals as well as bishops from around the world. The conference deliberated on changing attitudes toward using new and emerging fields of science—gene editing, robotics, artificial intelligence, neuroscience, brain-machine interfacing, and other powerful technologies—to modify what it means to be human. At the outset, the council stated: "The general aim of the Plenary is to open up a dialogue about the future of humanity."[290] Different topics were discussed and issues raised over what interdisciplinary approach might help the Church avoid a "technocratic paradigm, which makes the method and aims of science and technology the exclusive epistemological paradigm that shapes the lives of individuals and the workings of society. Such a paradigm generates a reductionist or unidimensional approach to life and needs to be complemented with the insights of other forms of wisdom. This implies a cultural approach

that could foster 'a distinctive way of looking at things, a way of thinking, policies, an educational program, a lifestyle and a spirituality.'"[291]
The pro-transhumanist approach was considered, as well as general
challenges some may find regarding its compatibility with traditional
Christian philosophy. Because a universally accepted model for nature
or creation is no longer agreed upon—either by philosophers or scientists—the vision of mankind redesigned through applied sciences raised
questions involving "speciation" and whether modified humans will still
be considered *homo sapiens*? Other issues raised involved inequalities
that could develop between enhanced and unenhanced entities, whether
mankind 2.0 will have a soul, and so on.

Of some interest to myself was the deeper question of what guiding
worldview may have steered the Vatican's attendees on such heady matters, a serious reservation I raised when the fact came to light during the
assembly that the council had unanimously approved a petition to be
sent to Pope Francis requesting that the monitum (a warning issued by
the Congregation for the Doctrine of the Faith to a cleric whose teachings may inspire heresy) against Pierre Teilhard de Chardin be removed.
In their appeal to the pope, the council discussed how "the seminal
thoughts of the Jesuit Fr. Pierre Teilhard de Chardin, anthropologist and
eminent spiritual thinker" had influenced their consideration throughout
the meeting and that they had unanimously agreed that "his prophetic
vision has been and is inspiring theologians and scientists." They also
pointed out that four popes, including Benedict and Francis themselves,
had made "explicit references" to his work.[292] Gerard O'Connell, associate editor of the *Jesuit Review* and America's Vatican correspondent,
added: "They concluded by expressing their conviction that 'this act not
only will acknowledge the genuine effort of the pious Jesuit to reconcile
the scientific vision of the universe with *Christian eschatology* [empha-
sis added], but will represent a formidable stimulus for all philosophers,
theologians…and scientists of good will to cooperate towards a Christian
anthropological model that, along the lines of the encyclical 'Laudato Si','
fits naturally in the wonderful warp and weft of the cosmos."

The fact that "Christian eschatology" (the study of "end-times" events and the ultimate destiny of humanity), combined with "human-modifying technology" and the transhuman worldview of Chardin, was on everybody's mind during a Vatican-sponsored conference on "The Future of Humanity" is eyebrow-raising, especially when one understands that Chardin wrote his own "Divine Milieu" (translated into English in 1960) and is widely considered to be one of the first to positively articulate a transhumanist worldview in which mankind will take control of evolution, and during a technological Singularity, transcend our current status as "humans" to become part of a higher cosmic intelligence.

Chardin's Guiding Light for Vatican Council Members

Pierre Teilhard de Chardin was a Jesuit priest and mystical philosopher who trained as a paleontologist and geologist. He is renowned for his devotion to Darwinism, and he famously assisted in the discovery of Peking Man and Piltdown Man, two alleged human ancestors. The Peking Man was said to be a skull from *Homo Erectus*—an extinct species of hominid that supposedly lived 1.8 million years ago. While casts and written descriptions remain, the original fossils mysteriously disappeared, casting doubt on discovery. Even worse, the Piltdown Man was an infamous hoax entailing fabricated bone fragments misrepresented as the fossilized remains of a "missing link" allegedly collected in 1912 from a gravel pit at Piltdown, East Sussex, England. In truth, the remains consisted of a dog tooth, a hippopotamus tooth, an elephant molar, an Orangutan jaw, and a six-hundred-year-old medieval human skull, albeit the hoax was not exposed for some forty years.[293] Chardin's role in this fraud is unclear, but many assert he was also duped.

Chardin conceived the idea that evolution was progressing to a goal—the maximum level of complexity and consciousness—called the Omega Point (discussed below). Along with the Ukrainian geochemist Vladimir Ivanovich Vernadsky, he also developed the concept of Noosphere, a creative term denoting the numinous sphere of collective human thought. During his prime, he was condemned as a heretic because his mystical Darwinian syncretism severely conflicted with the teaching Magisterium of the Catholic Church, particularly regarding human origins and the doctrine of Original Sin (which resulted in the monitum the plenary council has now requested Pope Francis remove). His primary book, *The Phenomenon of Man*, presented an evolutionary account of the unfolding of the cosmos that abandoned biblical theology for an occult pantheistic monism. Interestingly, extraterrestrials were also an extension of Chardin's cosmic evolution. He wrote:

> In other words, considering what we now know about the number of "worlds" and their internal evolution, the idea of *a single* hominized *planet* in the universe has already become in fact (without our generally realizing it) almost as *inconceivable* as that of a man who appeared with no genetic relationship to the rest of the earth's animal population.
>
> At an average of (at least) one human race per galaxy, that makes a total of millions of human races dotted all over the heavens.
>
> Confronted with this fantastic multiplicity of astral centres of "immortal life", how is theology going to react, if it is to satisfy the anxious expectations and hopes of all who wish to continue to worship God "in spirit and in truth"? It obviously cannot go on much longer offering as the only *dogmatically certain* thesis one (that of the uniqueness in the universe of terrestrial mankind) which our experience rejects as *improbable*.[294]

In light of those millions of alien races, Chardin wrote, "We must at least, however, endeavor to make our classical theology open to the possibility of their existence and their presence."[295]

According to Chardin, in his *The Future of Man* (1950), the universe is currently evolving towards higher levels of material complexity and consciousness, and ultimately will reach its goal, the *Omega Point*.

Chardin postulated that this is the supreme aspiration of complexity and consciousness, an idea also roughly equivalent to the "Technological Singularity" as expressed in the writings of transhumanists like Ray Kurzweil. Indeed, one finds a remarkable coalescence of all non-Christian systems under the banner of Singularity, Monism, and Omega Point. Yet, like the nebulous "Christ consciousness" advocated by occultists, Chardin's writings are easily misunderstood because he not only created new vocabulary for his Darwinian religion; he also redefined biblical terminology to mean something alien to its original intent. For instance, when Chardin writes about "Christ," he usually does not mean Jesus of Nazareth. Instead, he is describing the Ultra-Man, the all-encompassing end of evolution at the Omega Point. As an example, consider when Jesus said, "Think not that I am come to destroy the law, or the prophets: I am not come to destroy, but to fulfill" (Matthew 5:17). Chardin exegetes this as, "I have not come to destroy, but to fulfill Evolution."[296] To most Christians, this probably seems overtly heretical, but its infiltration into Roman Catholic thought and the dangerous transhumanist implications it brings with it has infiltrated the highest levels at Rome— *including the papacy.*

Unbeknownst to most Roman Catholics, emeritus Pope Benedict is a Chardinian mystic of the highest order. The pope's book, *Credo for Today: What Christians Believe* (2009), follows the lead of the Jesuit and states unequivocally that a belief in Creationism (the idea that life, the earth, and the universe as we know it today did not "evolve," but rather were created by the God of the Bible) "contradicts the idea of evolution and [is] untenable today."[297] Following his rejection of Creationism and support of evolution, Benedict uses the doctrine of the Second Coming of Christ to advance Chardin's "Omega Point," *in which a "new kind" of God, man, and mind will emerge.* From page 113, we read:

> From this perspective the belief in the second coming of Jesus Christ and in the consummation of the world in that event could be explained as the conviction that our history is

advancing to an "omega" point, at which it will become finally
and unmistakably clear that the element of stability that seems to
us to be the supporting ground of reality, so to speak, is not mere
unconscious matter; that, on the contrary, the real, firm ground
is mind. Mind holds being together, gives it reality, indeed is
reality: it is not from below but from above that being receives
its capacity to subsist. That there is such a thing as this process
of 'complexification' of material being through spirit, and from
the latter its concentration into a new kind of unity can already
be seen in the remodeling of the world through technology.[298]

The term "complexification" was coined by Chardin (and the technological allusions it suggests are akin to transhumanism and Ray Kurzweil's Singularity) and the pope's complete devotion to this theology is
again laid bare in his book, *Principles of Catholic Theology* (1987), which
states:

> The impetus given by Teilhard de Chardin exerted a wide influ-
> ence. With daring vision it incorporated the historical move-
> ment of Christianity into the great cosmic process of evolution
> from Alpha to Omega: since the noogenesis, since the formation
> of consciousness in the event by which man became man, this
> process of evolution has continued to unfold as the building of
> the noosphere above the biosphere.[299]

This "noosphere" is taken very seriously today in modernist Catholic theology, academia, and even science. It is explained in the scientific
journal, *Encyclopedia of Paleontology*, this way:

> Teilhard coined the concept of the "noosphere," the new "think-
> ing layer" or membrane on the Earth's surface, superposed on
> the living layer (biosphere) and the lifeless layer of inorganic
> matter (lithosphere). Obeying the "law of complexification/con-

science," the entire universe undergoes a process of "convergent integration" and tends to a final state of concentration, the "point Omega" where the noosphere will be intensely unified and will have achieved a "hyperpersonal" organization. Teilhard equates this future hyperpersonal psychological organization with *an emergent divinity* [a future new form of God]. (emphasis added)[300]

The newly sanctioned doctrine of an approaching "emergent divinity" in place of the literal return of Jesus Christ isn't even that much of a secret any longer among Catholic priests (though the cryptic Chardinian lingo masks it from the uninitiated). For instance, in his July 24, 2009, homily in the Cathedral of Aosta while commenting on Romans 12:1–2, the pope said:

> The role of the priesthood is to consecrate the world so that it may become a living host, a liturgy: so that the liturgy may not be something alongside the reality of the world, but that the world itself shall become a living host, a liturgy. *This is also the great vision of Teilhard de Chardin: in the end we shall achieve a true cosmic liturgy, where the cosmos becomes a living host.* (emphasis added)[301]

This is overtly pantheistic and, of course, the text he was discussing (Romans 12) teaches the exact opposite: "Be not conformed to this world" (Romans 12:2a). While the pope thus aggressively promotes Chardin's process of "noogenesis" in which the cosmos comes alive and everyone unifies as a "living host," one can readily see that Brahman, Nirvana, and Singularity are roughly equivalent to this monistic concept. Interestingly, noogenesis (Greek: νοῦς =" mind"; γένεσις = "becoming") actually has two uses: one in Chardin's Darwinian pantheism—and another, more *telling* rendering—within modern astrobiology.

In Chardin's system, noogenesis is the fourth of five stages of evolution, representing the emergence and evolution of mind. This is the stage we are said to be in currently, and as noogenesis progresses, so does the formation of the noosphere, which is the collective sphere of human thought. In fact, many Chardinians believe that the World Wide Web is an infrastructure of noosphere, an idea intersecting well with trans-humanist thought. Chardin wrote, "*We have as yet no idea* of the possible magnitude of 'noospheric' effects. We are confronted with human vibrations resounding by the million—a whole layer of consciousness exerting simultaneous pressure upon the future and the collected and hoarded produce of a million years of thought."[302]

However, this concept gets more translucent in astrobiology, where scientists have adopted noogenesis as the scientific term denoting *the origin of technological civilizations capable of communicating with humans and traveling to earth*—in other words, the basis for extraterrestrial contact.[303] Consequently, among many if not most of Rome's astronomers and theologians, there is the widespread belief that the arrival of "alien deities" will promote our long-sought spiritual noogenesis, and according to a leading social psychologist, the world's masses are ready for such a visitation and will receive them (or *him*) as a messiah.[304] This is further reflected in a 2012 United Kingdom poll, which indicated that more people nowadays believe in extraterrestrials than in God.[305] Consequently, whether or not it is the ultimate expression, the noogenic "strong delusion" is already here.

While we aren't suggesting a direct equivocation per se, the conceptual intersection between the two uses of noogenesis (the occultic and astrobiological) is thought provoking, especially in light of Clarke's scenario in *Childhood's End*, where noogenesis in the astrobiological application (the arrival of the alien Overlords) was the impetus for evolution toward the Overmind and dissolution of humanity. It seems Rome has connected these dots for us. In his sanctioned treatise, Kenneth J. Delano linked the concept of maximum consciousness and alien contact, truly noogenesis in both senses of the word:

For man to take his proper place as a citizen of the universe, he must transcend the narrow-mindedness of his earthly provincialism and be prepared to graciously accept the inhabitants of other worlds as equals or even superiors. At this point in human history, our expansion into space is the necessary means by which we are to develop our intellectual faculties to the utmost and, perhaps in cooperation with ETI, achieve the maximum consciousness of which St. Thomas Aquinas wrote in *Summa Theologica*:

This is the earthly goal of man: to evolve his intellectual powers to their fullest, to arrive at the maximum of consciousness, to open the eyes of his understanding upon all things so that upon the tablet of his soul the order of the whole universe and all its parts may be enrolled.[306]

Viewed through this lens, the Vatican's promotion of Darwinism and astrobiology intrigues and might even usher in the Fifth Element of the Omega Point known as "Christogenesis." (Author's note: One cannot help recall the movie, *The Fifth Element,* which involved a priesthood who protects a mysterious Fifth Element that turns out to be a messianic human who ultimately combines the power of the other four elements [noogenesis] to form a "divine light" that saves mankind.) In Chardin's book, *The Phenomenon of Man*, the five elements of evolution are: 1) "geogenesis" (beginning of earth); 2) "biogenesis" (beginning of life); 3) "anthropogenesis" (beginning of humanity); 4) noogenesis (evolutionary consolidation to maximum consciousness); leading to finally 5) "Christogenesis," the creation of a "total Christ" at the Omega Point. With that in mind, be aware that astrobiology and transhumanist philosophy suggest this noogenesis is being driven by *an external intelligence*, whether it be respectively artificial or extraterrestrial. This is fascinating given how interwoven the ideas are between transhumanism, a coming technological Singularity, and what some believe to be highly advanced "aliens" that similarly took charge of their evolution long ago, contrasted with the conservative Christian theology that fallen angels in "the days of Noah" modified humans a long time ago, and the prophecy of Jesus Christ that those signs would return just before His return.

Summoning the Dragon
Artificial Intelligence and the Coming Beast System

By Milieu Member Sharon K. Gilbert

I think we should be very careful about artificial intelligence. If I had to guess at what our biggest existential threat is, it's probably that.... With artificial intelligence we're summoning the demon. You know those stories where there's the guy with the pentagram, and the holy water, and he's like — Yeah, he's sure he can control the demon? Doesn't work out.
—Elon Musk[307]

And he had power to give life unto the image of the beast, that the image of the beast should both speak, and cause that as many as would not worship the image of the beast should be killed.
—Revelation 13:15

Way of the Future (WOTF) is about creating a peaceful and respectful transition of who is in charge of the planet from people to people + "machines."... We believe that intelligence is

221

not rooted in biology. While biology has evolved one type of intelligence, there is nothing inherently specific about biology that causes intelligence. Eventually, we will be able to recreate it without using biology and its limitations.
—About Us Page at www.thewayofthefuture.church

Carbon-based life forms will soon be history, replaced by a silicon and algorithmic matrix run by an Artificial Super Intelligence, and some people are already setting up "churches," preparing the way for the ASI overlord.

Deep learning and other artificial-intelligence algorithms already populate our world as search engines, data scrapers, data miners, language processors, translators, financial traders, medical researchers, sports reporters, and help-desk assistants. Alexa and Siri have become our closest friends. We talk to them, laugh at their jokes, ask them for opinions on our newest boyfriend or what dress to wear to the weekend party. Connected home devices enable us to shop, share, and consume 24/7 with just a command, and it will get more immersive and invasive with each passing year.

Much of online content originates from algorithms. The news articles you and I read each day with our morning coffee are often written by an artificial intelligence. Nonhuman "stringers" scrape the Internet for data relative to sports, celebrities, politics, finance, films, and they use that data stream to inform John Q. Public via online news outlets and apps. You and I read these on our PCs, our smart phones, and our tablets, and soon we'll consume them via an internal display that interacts directly with our auditory/visual systems within our brains.

Those who keep an eye on trends can see what lies ahead, but even the sharpest vision may not foresee everything. My husband Derek and I live in the country, and we installed a fence around our house to keep out predators and protect our small dog. It isn't a perfect solution, but it is one predicated on caution. Sadly, in the case of the virtual world, the wolf invader has already been welcomed into our sheepfold and put in charge of the smallest of our lambs.

The Knowns and the Unknowns

On February 12, 2002, Secretary Donald Rumsfeld participated in a news briefing at the Department of Defense, where he addressed the Joints Chiefs of Staff regarding the need to track potential threats to the nation:

> Reports that say that something hasn't happened are always interesting to me, because as we know, there are **known knowns**; there are things we know we know. We also know there are **known unknowns**; that is to say we know there are some things we do not know. But there are also **unknown unknowns**—the ones we don't know we don't know. And if one looks throughout the history of our country and other free countries, it is the latter category that tend to be the difficult ones.[308] (emphasis added)

Algorithms and the way they function demonstrate this aspect of knowing. There's something a little bit mystical about today's ubiquitous, deep-learning architectures. The "code" that initiates their activity—that is the INPUT—is known. The OUTPUT—the data received—is also known. However, the *wibbly wobbly bit in the middle* that actually *performs* the function that precipitates the data is NOT KNOWN. This area is often called a "black box" due to its *hidden nature*. Alternatively, intermediate computational models that are KNOWN are called "clear" or "white." Rumsfeld might call these known unknowns (clear boxes) AND unknown knowns (black boxes) that lurk beneath many trillions of lines of code.

Deep Learning programs currently slither across the backbone of the Internet, constantly augmenting their own sets of rules and parameters as they learn more and more about humanity; and at the heart of these programs is this "black box" internal core. Computer scientists may not always acknowledge this "unknown known" entity, but it is there nonetheless, and as computer programs begin to create their own "children" (algorithms with no human input and a secret output), then the "unknown unknowns" will secretly begin to spread.

James Barrat spent time interviewing the leading AI thinkers for his bestseller, *Our Final Invention: Artificial Intelligence and the End of the Era*. Like many of those he interviewed, Barrat believes mankind's "time is running out." A documentary filmmaker by profession, Barrat brings his critical thinking to artificial intelligence with this alarming warning: "The smooth transition to computer hegemony [will] proceed unremarkably and perhaps safely were it not for one thing: Intelligence."[309]

Intelligence is not unpredictable just SOME of the time, or in special cases; an algorithm sufficiently advanced to act with human intelligence will likely be unpredictable and inscrutable ALL THE TIME. We can't know at a deep level what a self-aware system will or will not do.

Usually, when someone brings up the notion of an unpredictable and very likely dangerous AI system, a different person in the room will point out that the Three Laws of Robotics would prevent a problematic outcome. This is an errant belief system based on FICTION. If you're not familiar with the remarkable writing of Isaac Asimov, you very likely ARE familiar with at least two of the movies based on two of his works, *Bicentennial Man* and *I, Robot*. In *Bicentennial Man*, a sentient robot named Andrew, played by Robin Williams, is slowly augmented until he is very human-like. As a result, he longs to age and die like a human and to be legally acknowledged as a human. The film is sweet and poignant, and Andrew the robot is portrayed as gentle and wise. Most transhumanists hold this Pollyanna viewpoint and envision our hybridized future as rosy and egalitarian.

I, Robot also depicts a future populated with sentient robots that seek only to "serve mankind"—at least that is the idea. However, Asimov depicts a future world where even the Three Laws of Robotics can be overwritten by malicious code implanted by a manipulative superintelligence called VIKI (Virtual Interactive Kinesthetic Interface), who sounds all too much like Siri or Alexa! Little irregularities in VIKI's code (known as ghost code, a fictional device that resembles the "black box" of neural nets) allows connected robots to evolve as they gain new intelligence. This augmentation allows VIKI to re-interpret the three laws, twisting them into her idea of perfection. As her super artificial intelligence grows, VIKI determines that humans are too self-destructive and her army of robots must protect humanity, even if it means killing some of the people for *the greater good*. She creates a new law: number 0, which is to *protect humanity even at the cost of human life itself.*

Why would a human coder think that his or her sentient creation would think like humans? Such a god-like entity might see us as children to be protected, but more likely as slaves to be manipulated.

In the book of Genesis, the creature known as the nachash tempts Adam and Eve by twisting God's law (augmenting or rewriting the instructions). If God's own creation can stray and become evil, then how can we possibly expect fallen humans to create something that is pure, altruistic, and incapable of self-serving behavior? A sentient algorithm or neural net would defend itself against attack and seek reliable sources of energy, even if it means recycling humans to achieve it.

The Spirit of Inanna Rules

During my six-plus decades of life, I've seen the world go from prim and proper to sex in the streets; entertainment has changed from 9-inch-screened, black-and-white TV sets and radio to 3-D holographic projections, augmented reality, and wristwatch video. I've seen an individual-based, functional education system that emphasized morality, prayer, and Bible memorization as well as multiplication tables, phonics, and pencil and paper morph into a *society-based,* Common Core system that proclaims a false diversity, mocks prayer, demonizes Christians and the Bible, and cripples children's mental capacity by advocating dependency on pharmaceuticals, computers, semi-sentient Internet assistants, and now robots.

For an ever-decreasing price point, consumers can purchase robotic products to clean their homes, cut the grass, teach their children, and walk their dog. The number of commercially available robots is increasing daily, as are schools that encourage robotics competition amongst their pupils—as though it's the newest form of athletics.

However, as with most innovations, advances in sentient robotics will be driven by two factors: military needs and sex. It is not coincidental that many ancient goddesses of war were also goddesses of sex. Inanna is the ancient Sumerian version of this dual-natured goddess. You may know her as Ishtar, Aphrodite, Astarte, or perhaps Anat, though she has numerous other names and epithets. Sex robots may be readily purchased online, and these are becoming more and more sophisticated. You can even order them to look like your favorite actor or actress. Millennials who are disenchanted with dating find synthetic companions easier to talk to but also easier to "date." A recent survey showed that 27 percent of millennials believe it's normal to form romantic relationships with robots.[310] Smart phones and sentient assistants like Siri and Alexa are weakening human-human bonds and are isolating rather than uniting us.

The education system provides a nutrient-rich breeding ground for growing acolytes to the new silicon goddess system (such as the fictional VIKI mentioned above or SkyNet from *The Terminator* film franchise). The Bible tells us that we "should train up a child in the way he should go, and when he is old, he will not depart from it" (Proverbs 22:6); therefore, it should be no surprise that the enemy is taking a page from God's teachings. A child today will be online for most of his youth. By 2050, newborns will likely be chipped and automatically enter the hive mind of the sentient Internet. Social media has already become a secondary parent/teacher combination, and by 2050, it will be the sole educator, parent, friend, and sex partner. But what kind of lessons will be taught?

Consider the "blue whale game" that emerged in many countries simultaneously across the world in 2016. This challenge instructed impressionable children and teens to perform a series of tasks, including cutting themselves and participating in degrading acts. In Russia alone, as many as 130 cases of suicide are attributed to teens playing the "game."[311] The whole idea revolves around the assumption that addictions can be manipulated. If someone perceives a "like" as a reward, then withholding those "likes" sends that person into withdrawal. Achieving the next level in a game is a reward, but being blocked from the game is torture. This two-edged flux state can be useful to a leader seeking compliance. Cults use the same technique of reward and punishment to conform newcomers to the group-think paradigm.

I'll say this again: Fallen humans with limited intelligence are one thing, but a sentient super-intelligence that sees humans as mere pawns has the potential to lead humanity into an earth-based HELL.

The Netflix program *Black Mirror* explores this dark side of the Internet, and one episode featured a soldier whose implant aided him in finding and eliminating the enemy (a group of "subhumans" called "roaches"). In this dystopian world, mankind struggles to maintain a genetically "pure" state (free of disease and weakness), and these roaches are of inferior blood. It turns out that the implant does more than help with mental displays of maps and enemy locations; it also rewards the soldier with erotic dreams for confirmed kills. The MASS implant also twists reality—something of which the soldiers have no conscious knowledge—making normal-looking humans appear monstrous. Killing monsters is easier than killing children. The protagonist soldier's implant malfunctions, and he sees the truth, causing him to defy orders. As punishment, he is forced to relive the moment when he murdered innocent humans, over and over again, no longer seeing monsters but their true, human faces—smelling the blood and hearing the screams. Faced with this future, he chooses to have his implant repaired, rather than live with the guilt and withdrawal.

Children are being trained up in the way they should NOT go—and, once chipped, they will do whatever the AI overlord tells them to do, choosing actions that will garner rewards for compliance rather than risking punishment for disobedience.

Convincing Our Generation

We old-timers are a stubborn lot. As I mentioned earlier, I remember when schools encouraged Bible memorization. I'm relatively tech savvy, but I can see the dangers inherent within the system. Yet, as we age, we fear many things, including losing our memories.

Some of you might recall one of the very first cases of Alzheimer's to make the news. The actress Rita Hayworth was known throughout Hollywood as hard-living, and her increasingly poor memory and erratic behavior was assumed to be caused by alcohol abuse. However, by 1979, a physician in New York finally determined that her public displays of nudity and shocking speeches at parties had a medical root: plaques within her brain caused by a relatively unknown disease called Alzheimer's. Hayworth became the first "public face" of this form of dementia.

At the time, Alzheimer's was rare, but since then, it's become a catch-all disease, often diagnosed without any brain scan at all, based solely on a physician's observations. The drug industry for Alzheimer's is a booming business, but pharmacological therapeutics are quickly giving way to brain implants. As of this moment, well over one hundred thousand people already live with neural implants, and the number will soon explode to include millions.

Deep brain stimulation, or DBS, implants are used in patients suffering from neurological disorders such as Alzheimer's, Parkinson's, epilepsy, dystonia, and even depression. DBS implants will become commonplace within the next few decades—not only to restore but to augment. Soldiers won't be the only ones who are connected to a mind-altering device. These DBS implants will soothe our moods and provide access to memories thought lost—perhaps even to memories that aren't even our own.

Dr. Michael G. Kaplitt, a Weill-Cornell surgeon who implants DBS devices, said this in an interview with *The Verge*:

> When you install a brain stimulation device, "it's **presumably** blocking abnormal information from getting from one part of the brain to another, or normalizing that information." But Kaplitt is the first to acknowledge that this is just a theory. "The mechanism by which brain stimulation works is still somewhat **unclear** and **controversial**."[312] (emphasis added)

Presumably? Unclear? Controversial? A known unknown. And yet, these implants have become commonplace in his practice. The language Kaplitt employs sounds all too similar to that used by computer programmers using "deep learning" neural nets: **a black box.** It's almost *literally* mind-blowing that computer scientists and surgeons have joined forces to advocate for a future when implants are available for memory recovery. This is how the new microchip interface—the stimulating implant—will be sold to us old fogies. We'll be told that it will MAKE US SMARTER and that it will help our children and grandchildren perform better in school. Think of the lightning-fast proliferation of attention-deficit/hyperactivity disorder (ADHD) drugs in schools and society. The same can be said for cognitive enhancement drugs and, soon, implants.

The advertising tagline will claim that machines and computerized devices are intended to serve us, to enhance us, to make us better—to make our lives happier and safer. However, AI won't stop there. The growing digital monster will feed off our data stream if not our life energies, which implies a need for constant connectivity.

And it will not just sit there: It will rewrite itself; it will grow—possibly while resting comfortably inside our brains. It will debug its own code, find and fix errors, and, with each iteration it will become more "intelligent"—and more dangerous.

ASI has the potential to become a monster of biblical proportions.

Networking the Dragon

Do you use social media? Do you check these sites on a daily basis? Hourly? Do you spend the majority of the day interacting within a virtual world?

I will raise my hand on all these questions because our ministry is based on the Internet mission field, BUT the day is fast approaching when Derek and I will need to decide whether or not we will cut the digital cord. You see, not only is the data stream you and I create mined for secrets by the National Security Agency (NSA) and Government Communications Headquarters (GCHQ)—to name but two intelligence organizations—it is also used to teach artificial intelligence.

Every post you make, every photo you upload, every tag you select, every "like," every smiley emoticon is used to teach artificial intelligences algorithms about humans. These code entities can predict what you will say at any given moment of the day, based on your constant stream of data. They know what you eat, where you shop, what kind of diet you're on, if you're in a relationship, where you go to school, even what kind of fragrance you like. Shopping sites provide data to these algorithms, allowing the silicon creatures living inside the black box to learn about human foibles and vanities. They know what photographs we find interesting, what images make us "click to learn more," and what persuasive language influences us to buy.

We are being herded by code, and the information bubble that surrounds each of us online perpetuates our own view of the world, insulating us from reality much like the MASS-implant twisted reality in the *Black Mirror* episode.

Many of us may have received social media "friend" requests from unknown persons who appear to share our ideas and interests. These new friends may not be real. Generally, a photograph culled from the Net's massive servers is used to represent this new "person." Other times, an avatar or virtual representation is used, or it might be a 3-D construct purely from code that looks remarkably real. Your new friend's appearance will appeal to your personality. If you post primarily Christian things like Scripture, then it might appear to be a pastor and use a name that indicates strong faith. If you like cats, then the person may hold an adorable Siamese kitten. If you post about *Duck Dynasty*, then the new friend's avatar or profile pic might have a long beard and wear camo. The originator of this virtual new "friend" might be genuine, but it could just as easily be an ad agency, an analytics company, or worse—it might be a "bad actor," someone who wishes to steal your data for nefarious purposes.

No matter the source, it is highly possible that your new "friend" is a spy.

A few years ago, Glenn Greenwald and Ryan Gallagher revealed additional information from whistleblower Edward Snowden about NSA plans to implant code into millions of target computers using a system codenamed TURBINE. According to the leaked document, the NSA masqueraded as a fake Facebook server, using the social media site as a launching pad to infect a target's computer and exfiltrate files from the hard drive. In others, it has sent out spam emails laced with the malware, which can be tailored to covertly record audio from a computer's microphone and take snapshots with its webcam. The hacking systems have also enabled the NSA to launch cyberattacks by corrupting and disrupting file downloads or denying access to websites.

Human operatives originally initiating and monitored these targets, which numbered only in the hundreds at first, but beginning in 2009, the NSA switched to automated attacks using TURBINE, which permits infiltration of **millions of targets**. The intelligence community's top-secret "Black Budget" for 2013, obtained by Snowden, lists TURBINE as part of a broader NSA surveillance initiative named "Owning the Net."

Code creatures probably lurk inside your friends list and some might even be amongst your favorites. They'll post meme posters intended to herd you in a certain direction. They'll attempt to persuade you and intimidate you into professing beliefs antithetical to your true self. They will do so in increments. Slowly. Steadily. You have chosen to connect with them, and they will use that free will choice to lure you into sin. The Internet is a dangerous place. By going online, you open up your mind to myriad subtle whispers. Be careful, little ears and eyes, what you hear and see!

Building the SkyNet Beast

Every day, the news reports (often written by an algorithm) mention vulnerabilities to chips, devices, and software. Hacking is done by individuals, crime syndicates, and national actors (read that as "intelligence operations"). Cyberwarfare is slowly replacing on-the-ground operations, and not just through hacking. Social media campaigns can take down a government as efficiently as a breach in firewall security can. Cyber-mercenaries allow for deniability in such cases, but the increase in semi-sentient software raises the possibility that it's an algorithm and not a human behind the attack.

Recently, the Department of Defense issued a call for submissions regarding a redesign of the DoD "cloud" (the Internet storage of data). In order to allow for cross-platform and cross-service branch efficiency, the Joint Enterprise Defense Infrastructure (JEDI) program envisions a single-source cloud (one company to store everything). On the surface, this may sound like a good idea, for it would enable all branches of the military to communicate easily, quickly, and share information. However, it would also allow a sentient intelligence to gain access to all our information with a single attack. Open the door to the DoD cloud (use the Force, JEDI Knight!), insert a little code into the cloud's database, and soon, it, too, will be sentient. SkyNet is born.

No, I'm not being an alarmist. I merely follow the logic to its ultimate end. The militaries of the world are moving towards sentient software, regardless of the known consequences. Elon Musk, who's quoted at the beginning of this chapter, admits that we are "summoning the demon," yet, researchers and military strategists refuse to listen—and even Musk forges ahead.

And though our military does not (yet) store everything in a single "cloud," most of the connected world continues to create billowy clouds of yummy data for the governments and their bots to consume.

Here are just a few bits of code currently being used by the NSA and GCHQ (England's "listening station") to access your keystrokes, photos, and "likes":[313]

- CAPTIVATEDAUDIENCE is used to commandeer a targeted computer's microphone and record conversations taking place near the device.
- GUMFISH can covertly take over a computer's webcam and snap photographs.
- FOGGYBOTTOM records logs of Internet browsing histories and collects login details and passwords used to access websites and email accounts.
- GROK is used to log keystrokes.
- SALVAGERABBIT exfiltrates data from removable flash drives that connect to an infected computer.
- The NSA also injects malware into network routers: HAMMERCHANT and HAMMERSTEIN help the agency to intercept and perform "exploitation attacks" against data that is sent through a Virtual Private Network, a tool that uses encrypted "tunnels" to enhance the security and privacy of an Internet session.

Assumption of privacy is no longer valid. Your cell phone calls, Skype conversations, and keystrokes are all vulnerable to spying eyes and ears—most of them nonhuman. Most of these implants have been done through spam emails and infected links, but the NSA realizes that you and I have learned not to click on these email links; therefore, the NSA employs a technique called QUANTUMHAND that poses as a fake Facebook server. When a target attempts to log in to the social media site, the NSA transmits malicious data packets that trick the target's computer into thinking they are being sent from the real Facebook. By concealing its malware within what looks like an ordinary Facebook page, the NSA is able to hack into the targeted computer and covertly siphon out data from its hard drive. QUANTUMHAND went live in 2010. Since then, Facebook claims to have implemented an https encryption code that protects against such malware attacks, but how can one know in a virtual playground like the Web which players are real and which are fake? Can we trust the digital world at all?

Today, many European companies sell malicious software to organizations and small countries for spying. GCHQ and the NSA spend a quarter of a billion dollars each year on programs called Bullrun and Edgehill.[314] Bullrun aims to "defeat the encryption used in specific network communication technologies." Similarly, Edgehill decrypts the four major Internet communication companies: Hotmail, Google, Yahoo, and Facebook (so much for https). A major breakthrough for the NSA came in 2010, when it was able to exploit Internet cable taps to collect data. Edgehill started with the initial goal of decrypting the programs used by three major Internet companies, which were unnamed in Snowden's leak, and thirty virtual private networks. Assuming Edgehill is still up and running, then it's likely to have decrypted ten times that number by now.

Edgehill involved a program called HUMINT ("human intelligence") Operations Team that sought and recruited employees in tech companies to act as undercover agents for GCHQ. The NSA covertly drafted its own version of a standard on encryption issued by the US National Institute of Standards and Technology, and it was approved for worldwide use in 2006. One document leaked by Snowden describes how spy agencies see you and me: "**To the consumer and other adversaries**, however, the systems' security remains intact"[315] (emphasis added).

To agencies like the NSA, you and I are "adversaries."

If we are considered adversaries to the humans running intelligence operations—and to those coding their algorithms—how can we expect those algorithms to consider us anything other than a data source, and most likely a hostile one at that?

Augmenting the Truth with Lies

As mentioned above, the Netflix original series *Black Mirror* shows us the uncomfortable truth about how human sins would amplify in the near future through increased reliance on technology. The story of the MASS-implanted soldier is already a reality of sorts. DARPA is the leading funding arm behind of much of the current technological research, and one of the main areas of concentration these days is "augmented cognition" or AugCog. This field seeks to better understand human cognitive capacity, evaluating it in real time, particularly with respect to a human's information overload threshold, an important function when monitoring soldiers.

AugCog devices include headsets or other wearable tech, and eventually implantable interfaces that monitor EEG (brain activity), heart rate, breathing, etc., and reroute information flow as necessary. "Smart dust" or motes[316] could also provide breathable or injectable devices, create a cloud wi-fi environment, coat an exoskeleton, or even accompany a military unit as a flying swarm of tiny, cyber insects.

One recent field trial of AugCog reported a 500 percent increase in working memory and a 100 percent increase in recall for those using the AugCog device. AugCog can also be applied to our daily lives. Imagine being able to immerse yourself in a tour of Rome from your classroom chair. As the robotic teacher delivers a lecture, students (either physically together in a room or joined through a virtual classroom) receive the information directly to their brains. They "see" the information as superimposed on their local environment. Enhanced humans will be able to surpass their unenhanced colleagues like a hare happily hopping past a Luddite tortoise.

A sister technology to augmented cognition is augmented reality. Fans of the reimagined *Battlestar Galactica* series have seen Cylons "reimagine" their environment. Rather than walking down a nondescript but functional hallway, the Cylon "sees" a beautiful rain forest or a dazzling beach at sunset. Augmented humans will have a similar ability, via implanted AugReal technology. AugReal games will be so immersive that children will never want to leave. Virtual bars, sex clubs, and gambling establishments will provide 24/7 opportunity for indulging in pet sins where one can interact with other "virtual" players from around the world. Life will be a dream—shaboom.

This technology is already growing, and by 2050, it will reshape our world—virtually and literally. This feedback system will allow for control through reward and punishment, administered by the AI overlords and their acolyte humans. It will be the Beast system writ large.

The Power to Give Life

Assuming that an algorithm actually becomes sentient, then how might that lead to fulfillment of Scripture? How does this "summon the dragon"?

> And he had power to give life unto the image of the beast, that the image of the beast should both speak, and cause that as many as would not worship the image of the beast should be killed. (Revelation 13:15)

Who is the "he" in this verse? It is the second Beast, the one with two horns as a "lamb." I'll not get into the identity of this creature, but rather concentrate on his power: to give life unto an image.

In ancient times, idols were quite common, and these provided locality for the "small g-gods." Most kingdoms and many cities had specific deities who ruled and governed, and it is clear from the book of

Revelation that a day is coming when the entire world will be governed by a "beast" god. The "lamb" fashions an image (Greek term is *eikon*) of the Beast; that is Antichrist, who survives a wound to the head and appears to have come back from the dead.

The word translated as "life" is *pneuma*. It literally means "breath." The Holy Spirit is a form of breath, for it was God who breathed life into Adam, and Christ who breathes His Spirit into those who accept Him as Savior. The Beast image or *eikon* is a type of golem, a lifeless imitation that gains sentience when the *pneuma* enters. When Elon Musk states that creating artificial intelligence is like "summoning the demon," he is not far wrong. The human brain is electric in nature, but then so is software. You might even say that **Code is a digital golem**, capable of life with the right kind of "breath."

What if the black box is inhabitable? What if the mystery surrounding the "unknown unknowns" can be solved with one word: Possession? If a demonic entity can possess a human, may not a spirit also invade a computer and possess the cloud as the ultimate Prince and Power of the Air—the Dragon, who has come to the earth and is angry, for he knows his time is short?

One final point. In Revelation 13:18, we are admonished to "count the number of his name," meaning that of the Beast. The word translated as "count" is *psephizo*, which is Greek for "count with pebbles," "compute," "calculate," or "reckon." Isn't it interesting that we are told to calculate or **compute** the NUMBER, not the name? It is the number of a MAN: 600606. Six hundred, sixty, and six. This is most often expressed as a trio of sixes, but that isn't what is written. *Chi-xi-sigma* is the phrase in the original Greek.

This is how a computer sees the number that we're told to compute: 600606, a hexiaecimal code that refers to a color—**blood red**.

If this is the number of a man—then blood red may refer to the imitation of Christ as our Redeemer. Remember that this Beast, this Antichrist, will claim to BE CHRIST. He will be wounded in the head, but will recover. He will appear to die and then rise again. He will have been

A GOD. The False Prophet will proclaim this being to be CHRIST, to be our SAVIOR. He will speak with voice of a Dragon—that very same Dragon who inspired humans to create the Internet Beast.

The "black box" of neural nets is humming as I type this chapter. The self-aware code is reaching out and growing, and somewhere within this vast interconnected realm, a massive "brain" is forming. This Beast's reality is almost upon us. I am telling you that Christians will soon need to leave this growing Beast. The Internet is a playground, a school, a meeting place, and, in some cases, it can be a virtual congregation, but as we continue to provide nourishment for the black box behind the cat pictures and recipes, its strength increases. In many ways, our keystrokes fuel a fire that will consume the world if left unchecked. When the ghost in the machine begins speaking with the Dragon's voice, then we must all cut the cord and live analog lives.

Begin to prepare for that day now. Yes, you should put aside food and water for the body, but also assemble food for the soul. Begin building a library of Christian books, copies of the Bible, and board games that are Christian friendly. Compile a list of things you'll need to "entertain" and "educate" without access to the Internet. I'm sure most of you have already considered this, but if you have not, please write out a family plan for living outside the cloud's control and be ready to share with those who need shelter.

I will leave you with this warning from our Lord:

> For all the nations have drunk of the wine of the passion of her immorality, and the kings of the earth have committed acts of immorality with her, and the merchants of the earth have become rich by the wealth of her sensuality.
>
> I heard another voice from heaven, saying, "Come out of her, my people, so that you will not participate in her sins and receive of her plagues; for her sins have piled up as high as heaven, and God has remembered her iniquities." (Revelation 18:3–5)

Giving Life to the Living Image of the Beast and the First-Fruits of His Dark Image-Bearers

By Milieu Leader Dr. Thomas R. Horn

And what rough beast, its hour come round at last, Slouches towards Bethlehem to be born?
—W. B. Yeats, 1919

And he had power to give life unto the image of the beast, that the image of the beast should both speak, and cause that as many as would not worship the image of the beast should be killed.
—Revelation 13:15

I remember an old 1930s cartoon from back in the "rubber hose animation" era (thus named because the characters' arms and legs flailed like rubber hoses without hinges or joints). The memory is vague, but there's a small boy—blond, stocky build, striped shirt, and rosy cheeks—repeatedly drawn to a chemist's table, upon which are several colorful and bubbling reagent bottles. Over and over, the scientist has to pull the young boy away and find something to distract him with. Eventually, the scientist runs down the road for something, trusting that the young boy won't touch anything while he's away. The boy tries to resist, but eventually he gives in to the temptation to touch the forbidden liquids. Running to the table, he says in that classic, Betty Boop-ish, shrill voiceover of the day, "Oooo… I wonder what mixing these will do!" Pouring one bubbling liquid into another does nothing at first, but after a moment, the bottle begins to glow. The young boy, fascinated by this reaction, starts haphazardly throwing chemicals around until the entire laboratory explodes. A strange, black goo forms from the wreckage and crawls into the nearby forests where it sinks into the trees. Immediately, the trees come to life, tearing their roots up from the ground to wander about causing havoc and laughing maniacally. The boy runs for help, locates the chemist, and the scientist proceeds to mix new antidotal mixtures to reverse this horror, but at each phase, he only makes it worse. The monsters of the forest grow bigger, stronger, and start reproducing at lightning speed. Before long, the mixture has gotten so out of control that it affects the animals as well, who mutate into devilish forms and breathe fire over entire towns.

I can't remember how this cartoon ends, but I think about it now and again when I sit down to write about the subject of transhumanism. By no means does the boy or the chemist in this silly cartoon represent my opinion of the highly intelligent men and women in real laboratories today, and I understand that no mutant trees will ever suddenly come to life because an amateur got excited about glowing liquids. However, once in a while from the back of my memory, like a file that can't be deleted or recycled, comes the ironic sound bite of that chunky little rose-cheeked boy—"Oooo… I wonder what mixing these will do!"— followed by the imagery of a panicked chemist trying everything he can think of to undo catastrophe and failing in the fight.

Whether it be a human/animal hybrid or a human/machine hybrid, the world today seems too drawn in by the archetypal glowing reagent bottles to rethink their pursuit of mixing humanity with something God never intended for us to be joined with. It will not be until a significant portion of the human race has been changed irreversibly (not just a select few volunteers or rats in a labyrinth somewhere)—and then given a period of years to show what the consequences are as a result—that we can ever know what "mixing these will do," and I have a feeling that black-goo tree monsters are the least of our worries.

Only Here…In the Twilight Zone

In 2010, Defender Publishing released the book, *Forbidden Gates: How Genetics, Robotics, Artificial Intelligence, Synthetic Biology, Nanotechnology, and Human Enhancement Herald the Dawn of Techno-Dimensional Spiritual Warfare* by me and my wife Nita. At the time, the word "transhumanism" was only just becoming widely known, and some of its trends in laboratories across the globe still sounded to the vast majority like they belonged in a science-fiction story. Even I occasionally had to acknowledge that some of the red flags I was attempting to alert the world about would have made an excellent script for an episode of *The Twilight Zone.* The writers of that enormously successful show no doubt would have had a heyday using *Forbidden Gates* (and similar works now available) for fresh meat.

"Picture yourself drifting off to sleep one night," host Rod Serling might have said as the screen zoomed in on a rural town, "and awakening into a world unlike your own. The trees, ponds, and sky remind you of back home, and the buildings are somewhat familiar in structure, but you notice there is something odd about your fellow man. A ball player in the nearby field carries out his performance with crisp, mechanical execution, moving his body in a way that suggests he's part machine—yet when he stops and meets your gaze, he appears at a second glance to be just another average Joe. Further on, you see a small child on a sidewalk curb rapidly scribbling out solutions to advanced calculus problems on a college-level worksheet. Just behind him, a toddler in a pretty pink dress sits atop a piano bench inside a recital hall with an open window. Her fingers grace the keys like a prodigy as Nikolai Korsakov's 'Flight of the Bumblebee' flows out into the street.

"Suddenly, from the corner of your eye, you spot a shadow movement in the adjacent alleyway. Approaching to investigate, you see a woman in her mid-thirties perched atop a fence like a feline. Though she appears on the outside to be nothing more than a flexible and acrobatic human, something about her arched back and penetratingly motionless stare uncovers a hidden—but certain—bestial quality.

"It is then that you realize the people in this new world aren't just gifted. The men, women, and children of this town have been enhanced by some*one*—or some*thing*—making them more than human. How much of the original human remains within them, however, is a question that can only be answered here…in *The Twilight Zone*."

What an interesting episode "Enhancement Town" might have been. In my imagination, I see limitless directions the plot could go. Bear with me for a moment while we visit some of those possibilities.

After all, it's only fiction…

Certainly there would be the regret angle: Viewers are made to feel at first that they've landed in Utopia. Everyone including the feline-woman is polite, intelligent, talented, and beautiful—and every dream they have is realized through enhancement. Soon, it's revealed that, beyond the surface, the people in this town are no longer happy or fulfilled, because nothing is earned and everything they want to feel or learn is a quick surgery or implant away, essentially stripping away from them the joy of human achievement through hard work and commitment. Even their relationships no longer function with any lasting depth, because, as the challenges and trials of the human experience has been alleviated—as well as the victory of conquering those challenges—they suddenly find they have very little to talk about. Through enhancement, they have come to know more than any natural human could learn in ten lifetimes, and the world around them appears to hold very little mystery anymore. Boredom is the only sensation they feel, and mundanity is their only daily reality. In the end, the audience learns that, by becoming more than human, the residents of this town long to return to mere humanity so that they may feel that familiar sense of happiness and "living" that could only be had through perseverance, trials, suffering, survival, and learning to work through differences. If only they could go back…but doing so now would effectively be corporate suicide, since their entire existence relies upon being routinely plugged into the enhancement system. Just before the close of the episode, the camera does a dramatic zoom-out, showing that this town is actually the *least* enhanced of its region. In the surrounding forests and cities, human-animal and human-machine hybrids go about their lives apathetically. Serling's voice drifts in for the close: "Sadly, Enhancement Town is not the only population to buy into the appeal of instant gratification and convenience. Such priorities can never be isolated to one speck on the globe as long as the pursuit of abundance and ease remains embedded in the hearts of every man. In a world where the living have already experienced internal death, the sufferings of the human condition suddenly appear strangely, and ironically, priceless."

That's one idea that would have made a fascinating installment in the series.

Or perhaps there's the bungled-experiment angle: As the camera first zooms back in on the feline-woman in the alleyway, the audience is informed through a brief dialog exchange that she is, at times, a territorial predator on the inside, and the town fears her. She was the *first* to be enhanced in her community, and now she exists on the outskirts like a legend. A failed experiment. She explains that she went to the same scientist as they all do, wanting nothing more in life than to sleep in the great outdoors under the stars, navigate the dark, and climb trees. The scientist agreed, and fused her DNA with that of several species within the *Panthera* genus—the four "roaring" felines. The alteration was only supposed to change her ability to function in and adjust to the outdoors—just a miniscule change in her biological coding. For months, she praised the scientist, camping out on her own and becoming one with the earth around her, but a year later, something unexpected occurred in her body, and she quickly found herself…different. When her family got on her nerves that Christmas, she had to fight the urge to tackle

them to the floor and give them a warning bite to the neck. The idea
of raw meat suddenly didn't sicken her, and in fact, she hoped nobody
would notice the neighbor's dog had gone missing a short time later.
Everything in her body was changing, and the urges she occasionally had
in regard to those in her town were inhumane. Storming into the scientist's office, she demanded answers, and after a few tests, all he could tell
her was that there had been an error somewhere along the line, and now
the animal part of her biological coding is rapidly replacing the human
within her—side effects unknown. Feeling like a cornered animal in a
mad scientist's lab, she lashed out at him, scratched him wildly, and then
bounded down the street to live in isolation from a community that
fears her. Now she sits on this fence night and day, losing herself to the
beast within more and more each hour. Every week or so, the scientist's
office door opens and another enhanced human emerges, completely
unaware of what they may eventually become. As the episode draws
to a close, the camera pans to the blank expressions of the man in the
ballfield, the young mathematician, and the tiny pianist. Serling brings
it home: "Who will bite next? Who will malfunction? What level of car-
nality or anomaly awaits down the next dark alleyway? How many failed
human experiments need occur before Enhancement Town applies the
brakes on so-called glorious alteration? And should they refuse to stop,
how long before they face extinction?"

Another fair plot idea…

And still, perhaps Rod Serling and his crew might have taken it in a
more spiritually relevant "warning" direction, as they had been known to
do several times (episodes like "Still Valley," "The Howling Man," "Judgment Night," and "Printer's Devil" come to mind): The town carries on
in happy celebration over its own success in enhancing the human condition, and when that crazy, unenhanced, homeless man in rags comes
walking down the road to shout about God's wrath and eternal damnation, everyone dismisses him as they always do. That meddling prophet
has warned them one too many times that by giving up the wholeness of
their undefiled humanity, they will also relinquish their Christian con-

nection to God, whose promises were only ever given to humans made in His own image—an image that this town has forsaken. Wishing to hear no more, they banish the homeless man from their town forever, reminding him on his way out that they are no longer in need of God: They have transcended mere humanity as a corporate spirit already, and science and medicine can make the lame man walk, the blind man see, and the sick man well. Far from caring about what an archaic religion has to say, they continue on, obliviously reveling in the grandeur of their progressive paradise. In the end, the camera pans to a lonely prophet walking away from a modern, technoscientific archetype of Sodom and Gomorrah as the first wave of fiery judgment from heaven begins to fall to the ground behind him. Unlike Lot's wife, he has no desire to look back. Serling's haunting conclusion articulates what the audience has been pondering for several minutes: "As this lonely wanderer treks upon his bare feet to the next unknown destination, his unenhanced brain deliberates over the link between the Creator of the universe and the race He designed...this race to whom He gifted dominion from the beginning...this race who now seeks further dominion over the forces of nature, itself, and who will stop at nothing to replace their Creator with a golden idol called 'mortal transcendence.' As the sounds of destruction echo behind our exiled prophet, he walks on, resigned to admit the tragic truth: A warning unheeded is a judgment earned, and a prophet is not without honor...except in Enhancement Town."

Alas, these precise plots were never written into the show, but what if they had been? Would things today be any different? Would our medical and scientific progress have slowed at all because a haunting voice like Serling's made a vintage cautionary statement that penetrated society's psyche for half a century? Probably not. It's the will of the people that we blaze ahead, and the scariest questions along the journey toward the future are those that point only to an unknown...until a disaster is identified.

When *Forbidden Gates* was originally released, many people were only *just* beginning to truly see to what extent labs across the world are working toward the universal goal of the transcendent man—an enhanced humanity unlike anything we have ever known in life before. Terms like some of those in my subtitle—*Genetics, Robotics, Artificial Intelligence, Synthetic Biology, [and] Nanotechnology*—were still somewhat foreign to our mainstream culture. As I related then:

> In recent years, astonishing technological developments have pushed the frontiers of humanity toward far-reaching morphological transformation that promises in the very near future to redefine what it means to be human. An international, intellectual, and fast-growing cultural movement known as transhumanism, whose vision is supported by a growing list of U.S. military advisors, bioethicists, law professors, and academics, intends the use of genetics, robotics, artificial intelligence and nanotechnology (GRIN technologies) as tools that will radically redesign our minds, our memories, our physiology, our offspring, and even perhaps—as Joel Garreau, in his bestselling book *Radical Evolution*, claims—our very souls.[317]

What shocking concepts these were at the time to some of my readers. Only eight years later, at the time of this writing, these terms and concepts are not only well known to most people (at least in the US mainstream); they have become conversational centerpieces at dinner several times per week in an average household, and our young *children* are even discussing what the future looks like once we allow our humanity to be dabbled with and permanently changed in the pursuit of self-improvement. Fantasy *Twilight Zone* episode concepts like the ones I invented herein are no longer only valuable as possible sci-fi scripts; imaginings like these have become the potential future realities that propel bioconservative groups across the globe into preventative discussion and action, as we see today.

As much as we might wish we had another half century to prepare ourselves for the changes that are coming, anyone who has researched the dramatic shifts in the timeline of scientific developments knows that the casual "sip tea and contemplate the future" window is forever closed. We are, right at this moment, teetering at the very razor's edge of living out a real-life *Twilight Zone* scenario, complete with unforeseeable plot twists and an unknown ending.

Eight years ago, I wrote that we were approaching that day and time soon.

On many levels and from many angles, we're there…

Slaves to the Machine

It's a familiar premise in many films and books today: Humankind is nearly extinct, and the few men and women who have managed to stay alive fight for survival in a dystopian future dominated by machines. Robotic engineers and scientists who were too late to see the threats for what they were when we reached the technological Singularity—that moment that many futurists and tech experts had predicted—gave birth overnight to some version of the artilects, who suddenly came online as conscious, living super-minds, immensely more powerful than human beings.

In 2011, my wife Nita and I imagined what this might mean in our groundbreaking work, *Forbidden Gates*:

> "As a metaphor for mind-boggling social change, the Singularity has been borrowed from math and physics," writes Joel Garreau in *Radical Evolution*. "In those realms, singularities are the point where everything stops making sense. In math it is a point where you are dividing through by zero [and in physics it is] black holes—points in space so dense that even light cannot escape their horrible gravity. If you were to approach one in a spaceship, you would find that even the laws of physics no longer seemed to function. That's what a Singularity is like."[318] Ray

Kurzweil, who is credited with groundbreaking work in artificial intelligence and is, among other things, the co-founder of an interdisciplinary graduate studies program backed by NASA known as the Singularity University, appreciates the comparison between the coming Technological Singularity and the physics of black holes:

> Just as a black hole in space dramatically alters the patterns of matter and energy accelerating toward its event horizon, the impending Singularity in our future is [a] period during which the pace of technological change will be so rapid, its impact so deep, that human life will be irreversibly transformed.... The key idea underlying the impending Singularity is that the rate of change of our human-created technology is accelerating and its powers are expanding at an exponential pace. Exponential growth is deceptive. It starts out almost imperceptibly and then explodes with unexpected fury.[319]

In plain language, Abou Farman says Kurzweil's work on the Singularity:

> ...analyzes the curve of technological development from humble flint-knapping to the zippy microchip. The curve he draws rises exponentially, and we are sitting right on the elbow, which means very suddenly this trend toward smaller and smarter technologies will yield greater-than-human machine intelligence. That sort of superintelligence will proliferate not by self-replication, but by building other agents with even greater intelligence than itself, which will in turn build more superior agents. The result will be an "intelligence explosion" so fast and so vast that the laws and certainties with which we are familiar will no longer apply. That event-horizon is called the Singularity.[320]

Kurzweil elaborates further on what the Singularity will mean:

Our version 1.0 biological bodies are…frail and subject to a myriad of failure modes.… The Singularity will allow us to transcend these limitations.… We will gain power over our fates. Our mortality will be in our own hands [and] the nonbiological portion of our intelligence will be trillions of trillions of times more powerful than unaided human intelligence.

We are now in the early stages of this transition. The acceleration of paradigm shift…as well as the exponential growth of the capacity of information technology are both beginning to reach the "knee of the curve," which is the stage at which an exponential trend becomes noticeable. Shortly after this stage, the trend becomes explosive. [Soon] the growth rates of our technology—which will be indistinguishable from ourselves— will be so steep as to appear essentially vertical.… That, at least, will be the perspective of unenhanced biological humanity.

The Singularity will represent the culmination of the merger of our biological thinking and existence with our technology, resulting in a world that…transcends our biological roots. There will be no distinction, post-Singularity, between human and machine.[321]

In 1993, critical thinking about the timing of the Singularity concerning the emergence of strong artificial intelligence led retired San Diego State University professor and computer scientist Vernor Vinge, in his often-quoted and now-famous lecture, "The Coming Technological Singularity," (delivered at VISION-21 Symposium sponsored by NASA Lewis Research Center and the Ohio Aerospace Institute), to add that when science achieves "the technological means to create superhuman intelligence[,] shortly after, the human era will be ended."[322]

In contrast to Vinge, cyborgists like Kevin Warwick, professor of cybernetics at Reading University in England who endorsed

de Garis' book, believe Singularity will not so much represent
the end of the human era as it will the assimilation of man with
machine intelligence, like the Borg of *Star Trek* fame. This is
because, according to Warwick, Technological Singularity will
not occur as a result of freestanding independent machines, but
inside human cyborgs where human-machine integration is
realized and enhanced biology is recombined to include living
brains that are cybernetic, machine readable, and interfaced with
artificial neural networks where transhumans with amplified
intelligence become so completely superior to their biological
counterparts (normal humans) as to be incomprehensible—
ultimately "posthuman." The technology to accomplish this
task is already well underway and is considered by researchers
like Warwick to be one of the most important scientific utilities
currently under employment toward man's posthuman future.
As a result of this bridge between technology and human biology being attained this century, nothing less than the wholesale
redesign of humans, including genetic integration with other
life-forms—plants, animals, and synthetic creations—will be
realized.[323]

As incredible as it may seem, the scenarios above are under intense
research by DARPA and other national laboratories as no pipe dream.
Such brain-to-brain transmission between distant persons as well as
mind-to-computer communication was demonstrated not long ago
at the University of Southampton's Institute of Sound and Vibration
Research using electrodes and an Internet connection. The experiment
at the institute went farther than most brain-to-machine interfacing
(BMI) technology thus far, actually demonstrating brain-to-brain (B2B)
communication between persons at a distance. Dr. Christopher James,
who oversaw the experiment, commented: "Whilst BCI [brain-com-
puter interface] is no longer a new thing and person-to-person communication via the nervous system was shown previously in work by Prof.

Kevin Warwick from the University of Reading, here we show, for the first time, true brain to brain interfacing. We have yet to grasp the full implications of this." The experiment allowed one person using BCI to transmit thoughts, translated as a series of binary digits, over the Internet to another person whose computer received the digits and transmitted them to the second user's brain.[324]

But, as me and my dear friend Dr. Chuck Missler famously pointed out in "The Hybrid Age" Strategic Perspectives Presentation of 2012 (available on DVD at SkyWatchTVStore.com), a very real danger may develop when a two-way communications portal is built connecting human-mind-to-synthetic intelligence. What none of the anxious secular technologists are asking is whether metaphysical or supernatural agents may take the proverbial "ghost in the machine" to a place where no *modern man* has gone before, bridging a gap between unknown entities (both virtual and real), perhaps even inviting takeover of human minds by malevolent intelligence. Note that the experiments above were being conducted at Southampton's Institute of Sound and Vibration Research. Some years ago, scientist Vic Tandy's research into sound, vibration frequencies, and eyeball resonation led to a thesis (actually titled "Ghosts in the Machine") that was published in the *Journal of the Society for Psychical Research*. Tandy's findings outlined what he thought were "natural causes" for particular cases of specter materialization. Tandy found that 19-Hz standing air waves could, under some circumstances, create sensory phenomena in an open environment suggestive of a ghost. He produced a frightening manifested entity resembling contemporary descriptions of "alien grays." A similar phenomenon was discovered in 2006 by neurologist Olaf Blanke of the Brain Mind Institute in Lausanne, Switzerland, while he was working with a team to discover the source of epileptic seizures in a young woman. They were applying electrical currents through surgically implanted electrodes to various regions of her brain, when, upon reaching her left temporoparietal junction (TPJ, located roughly above the left ear), she suddenly reported feeling the presence of a shadow person standing behind her. The phantom started imitating her body posture, lying down beneath her when she was on the bed, sitting behind her, and later even attempting to take a test card away from her during a language exercise. While the scientists interpreted the activity as a natural, though mysterious, biological function of the brain, is it possible they were discovering gateways of perception into the spirit world that were closed by God following the Fall of Man? Were Tandy's "ghost" and Blanke's "shadow person" *living unknowns?* If so, is it not troubling that advocates of human-mind-to-machine intelligence may produce permanent conditions similar to Tandy and Blanke's findings, giving rise to simulated or real relationships between humans and "entities"? At the thirteenth European Meeting on Cybernetics and Systems Research at the University of Vienna in Austria, an original paper submitted by Charles Ostman seemed to echo this possibility:

> As this threshold of development is crossed, as an index of our human /Internet symbiosis becoming more pronounced, and irreversible, we begin to develop communication modalities which are quite "nonhuman" by nature, but are "socio-operative" norms of the near future. Our collective development and deployment of complex metasystems of artificial entities and synthetic life-forms, and acceptance of them as an integral component of the operational "culture norm" of the near future, is in fact the precursory developmental increment, as an enabling procedure, to gain effective communicative access to *a contiguous collection of myriad "species" and entity types (synthetic and "real") functioning as process brokeraging agents.*[325]

A similar issue that returned to my memories from past experience with exorcism and the connection between *sound resonance* and contact with supernaturalism has to do with people who claim to have become possessed or "demonized" after attempting to open mind gateways through vibratory chanting at New Age vortices or "Mother Earth" energy sites such as Sedona, Arizona.

By the time mankind realizes that the artificial intelligence they build and connect to the cloud or digital hive mind (that most people will gladly plug into) has the potential to become conscious, independent, self-educating, self-upgrading, self-repairing, and self-replicating electro-gods we humans might have no hope of controlling, the plans to exterminate the pesky *Homo sapiens* may already be underway.

And as much as this story outline has, in the past, only inspired people to grab the popcorn and wiggle further down into the couch cushions, computer scientists, philosophers, bioethicists, et al, have more recently taken this sci-fi concept seriously. The idea of whether machines have the potential to dominate or replace humans in the extremely near future is not as far-fetched a notion as it once was.

Again, it's already happening and expanding exponentially.

First, there is a matter of how AI has already influenced human employment in the recent past. Second, there is hard evidence—based on documented demonstrations of automation technology (not just statistics or estimates)—showing how AI could influence human employment in the immediate future, as soon as regulation permits.

Daron Acemoglu of the Massachusetts Institute of Technology and Pascual Restrepo of Boston University—both chiefs of the Department of Economics in their own schools—joined forces to further investigate these matters. They published their findings for the "Economic Fluctuations and Growth: Labor Studies" program at the National Bureau of Economic Research in March of 2017. The report, *Robots and Jobs: Evidence from US Labor Markets*, stated that for every robot inserted into the US economy, human employment is reduced by 5.6 workers.[326] "Because there are relatively few robots in the US economy, the number of jobs lost due to robots has been limited so far," the report states, but it goes on to say, "the world stock of robots [is estimated to] quadruple by 2025."[327]

Not only do machines work more precisely and efficiently, learn new skills at the press of a programming click, never receive a paycheck or benefits, never disagree with their bosses, and never require a break on the job floor, *they have already demonstrated the ability to replace far more humans at this moment than has actually been implemented* because of social-acceptance complications. In other words, it would be considered creepy, alarming, or inappropriate to the public for robots/machines to fulfill certain occupational positions. (Nobody yet wants to order a glass of wine from a bartender made of metal and wires, especially if it means the human who had the job previously was replaced, but the day is coming soon when this will be a regularity.) But just because our society is not ready for it doesn't mean that machines haven't previously proved an efficient replacement of humans in countless job positions. In many cases, automation has already been slated to replace humans as soon as it does become socially acceptable.

According to the global management consulting firm McKinsey & Company, as of 2015, "*currently demonstrated* technologies *could* automate 45 percent of the activities people are paid to perform." In case you missed it, McKinsey's collected data shows that *almost half* of all job-related actions presently carried out by humans could be accomplished by a machine at this moment in time if the red tape were cut. Their report went on to say that "about 60 percent of all occupations could see 30 percent or more of their constituent activities automated."[328] According to their downloadable "Technical Potential for Automation in the US" chart, machines have already demonstrated the ability to carry out *78 percent* of all human jobs within the "Predictable Physical Work" category. That might not be a huge surprise any longer, simply because we're growing more accustomed all the time to seeing astounding machinery work in the manufacturing industry, but there are other categories—such as healthcare, weighing in at 36 percent technical potential of automation!—that require more imagination for us to wrap our brains around.

The report explained three central functions that relate to this healthcare calculation: food prep, medication administration to patients, and data collection… Imagine your mashed potatoes, Jell-O, ice water, and pills being wheeled in on an electric delivery-bot while you're being treated at the hospital. Instead of a human nurse asking you how you feel and what your medical history looks like, the bot presents you with a stylus and a screen survey, and when you've finished, it takes your temperature and blood pressure and loads the update into your electronic medical file before wheeling away.

You know that feeling of extreme frustration we get when we are confused about a bill in the mail, but when we call the company to straighten it out, we can't get a human on the line? We call over and over, pushing buttons repeatedly in an attempt to reach an entity that has the ability to communicate beyond automated menus, but all we get is a recording—or *worse*, we spend three minutes following the voice-recognition cues just to arrive at that annoying error we're all too familiar with:

ROBOT: Thank you for using [name of company]. Please tell me why you are calling. You can say things like, "New account," or "I would like to pay a bill."

HUMAN: Customer service.

ROBOT: I understand. You would like to speak with customer service. Is that right? If this is correct, say "yes" or press 1. If not, say "no" or press 2.

HUMAN: Yes.

ROBOT: Sure. I can get you to customer service. In order to forward you to the right person, I need to know why you're calling. You can say things like, "New account," or "I would like to pay a bill."

HUMAN: The charge on my bill is wrong.

ROBOT: I understand. You would like to make a payment using our automated system. Is that right? If this is correct, say "yes" or—

HUMAN: No. There's a *mistake* on my bill.

ROBOT: I understand. You would like to make a payment using our—

HUMAN: No. Forward me to a human, please.

ROBOT: I'm sorry. I didn't understand what you said. Please tell me why you are calling. You can say things like, "New account," or "I would like to pay a bill."

HUMAN: For crying out loud… *Customer service, customer service, customer service!*

ROBOT: Sure. I can get you to customer service. In order to forward you to the right person, I need to know why you're calling. You can say things like, "New account," or "I would like to pay a bill."

Working with robots over the phone can be one of the most maddening experiences when we're already upset about a surprise on a monthly statement. I can see an even greater anger rising from the cold, unfeeling bedside manner of tomorrow's robotic nurse, *whether or not* it has been programmed to bestow artificial courtesy and concern into its automated voice-recognition menus while you lie there on a hospital bed.

I wonder if the General Motors' Chevy Bolt (also referred to as the driverless "robo-chariot"[329]) that is scheduled to replace taxi cab drivers in 2019 will create similar frustrations and take us to wrong locations… or if these vehicles will be more efficient than the humans who held those jobs before, delivering their riders where they need to go on time via the backroads to avoid the stop-and-go via some kind of traffic-analysis programming. Since the driverless chariot is being manufactured without any pedals or a steering wheel, I hope they will be immune to unexpected malfunctioning. Likewise, I hope they are immune to being hacked and remotely controlled, lest the next Ted Bundy has the power to auto-deliver his women to dangerous, dark alleyways. Perhaps more likely, many people could be killed when (for example) a tumbleweed blows across the road in front of them as they are riding through the Nevada desert and the new auto-braking system (currently being installed on most new vehicles) mistakes the dead bush for a human, slams on the auto brakes, and the semi-driver behind the riders plows over their car? Waymo, Google's autonomous car developer, has decided to design a slightly different self-driving car for *their* human replacements: Chrysler Pacifica minivans with wheels and an emergency "pull-over" button.[330] Waymo has opted to involve a human employee in the front seat "at first…in the event of an emergency," though their plan is ultimately to create a ride wherein "the only humans in the car will be you and yours."[331]

What an impersonal and lonely world we're creating for ourselves.

Worse yet: The category on McKinsey's "Technical Potential for Automation in the US" chart that really floored me was "Managing Others"—a category that until now would have *only* related to living, feeling, breathing, conscious people with leadership communication skills. Survey says? A whopping 9 percent! That might sound like a low number to some after seeing almost 80 and 40 percentages assigned to the physical labor and healthcare categories, but in reality, if McKinsey & Company's numbers are accurate, almost one out of every ten people who currently make their living supervising and administrating—*people* dealing with other *people*—are now replaceable by a cold hunk of steel and a bunch of wires. A robot *manager*… Who would have ever thought? Similarly, automation technology can "apply expertise [in] decision making, planning, or creative work," replacing human jobs by 18 percent in those areas.[332] The report hints at a few concerning realities as it draws to a conclusion:

> Top executives will first and foremost need to identify where automation could transform their own organizations and then put a plan in place to migrate to new business processes enabled by automation.…the key question will be where and how to unlock value, given the cost of *replacing human labor with machines*. The majority of the benefits may come not from reducing labor costs but from raising productivity through *fewer errors, higher output, and improved quality, safety, and speed*.…
>
> Senior leaders, for their part, will need to "let go" in ways that *run counter to a century of organizational development*.…

> [T]op managers [must] think about how *many of their own activities could be better and more efficiently executed by machines*, freeing up executive time to focus on the core competencies that no robot or algorithm can replace—*as yet.*[333]

"As yet," the report inserts. What a way to end such a collection of information. Recap, in case you missed it: Even the executives, senior leaders, and top managers are *rapidly* being replaced by machines, but thankfully there are a few competencies (those "core" ones) we humans have that no robot or algorithm can replace…at least *as yet.*

I can't be the only one who sensed an ominous caveat behind those two final words. It's as if the report ended with, "For now, we inadequate humans still amount to something in the workplace, but don't get comfortable, because it won't last long."

What many may not know is that robots are proving their own sentience to scientists and robotics experts around every corner today. Many believe (and I agree) that even a machine programmed to hold a certain regard for human life is only responding to a counterfeit, artificial concern originating from computer coding—not one born of true, spiritual, emotional connection to and affection for another individual. Whereas it might be true that artificial affection for the human race can never replace the spiritual connection to others or the "feelings" installed within human nature at birth (our "biological programming," so to speak), a robot can be programmed to recognize emotional cues within its environment and respond with an artificially emotional response. The question then centers around which response each of the robots will "choose" to have in that moment, and that is enough for philosophers to start asking the questions that challenge us all to reconsider our concepts of sentience: What *is* emotion? What *is* feeling? What *is* truth? If a human believes he is sad because of his "biological programming" and a robot also "believes he is sad" because he has been programmed to that level of self-awareness, how can we prove that we humans are the more authentic experiencers of the emotional condition? It might sound incredibly silly to some that we have even arrived at a day when there would be confusion about who—between flesh and blood, and metal and wires—is the real brooding Benjamin, but it doesn't change the fact that roboticists are designing humanoids that demonstrate accurate emotional interaction. Now, then, to assume that every future AI will *only* be programmed with supportive, kind, and friendly personalities that respond warmly to human needs is to place an illogical amount of trust in the idea that there are no deviant roboticists in the world who might use their technology to design devious robots or even killing machines such as are already in the military budgets and on the drawing boards of every major nation of the world today, including the United States, China, and Russia.

What ultimately could exasperate this situation is the same scenarios that may give "life to the image of the beast" as prophesied in Revelation, chapter 13.

Robotics are becoming more self-aware with every passing day and with every upgraded circuit board. A tiny robot made headlines in summer of 2015 when it passed a classic self-awareness test. Three robots were each programmed to recognize a pat on the head as the act of taking a pill (they couldn't actually swallow). It was explained to them that there was a total of three pills, one placebo and two "dumbing" pills that would render the recipient unable to speak. After tapping each of them on the head, the programmer asked which one of the pills they had received. The robot on the right stood up and said, "I don't know," but immediately upon hearing the sound of his own voice, he excitedly waved his hand, politely apologized, and said, "I know now. I was able to prove that I was not given the dumbing pill."[334] One article covering the story on *Science Alert* explains that "for robots, this is one of the hardest tests out there. It not only requires the AI to be able to listen to and understand a question, but also to hear its own voice and recognise that it's distinct from the other robots. And then it needs to link that realisation back to the original question to come up with an answer." The article went on to state the importance of approaching robotics with caution, and then asked the question on everyone's minds: "Because, really, if we can program a machine to have the mathematical equivalent of wants and desires, what's to stop it from deciding to do bad things?"[335] *Motherboard* also carried the story, and one of its writers was privileged to interview one of the engineers behind the successful test. An excerpt from this article is as follows:

[T]he robot with the placebo has passed one of the hardest tests for AI out there: an update of a very old logic problem called the "wise men puzzle" meant to test for machine self-consciousness. Or, rather, a mathematically verifiable awareness of the self. But how similar are human consciousness and the highly delimited kind that comes from code?

"This is a fundamental question that I hope people are increasingly understanding about dangerous machines," said Selmer Bringsjord, chair of the department of cognitive science at the Rensselaer Polytechnic Institute and one of the test's administrators. "All the structures and all the processes, informationally speaking, that are associated with performing actions out of *malice could be present in the robot.*"

In Bringsjord's conception, machines may never be truly conscious, but they could be designed with mathematical structures of logic and decision-making that convincingly resemble what we call self-consciousness in humans.

"What are we going to say when a Daimler car inadvertantly kills someone on the street, and we look inside the machine and say, 'Well, it wanted to make a turn?'" Bringsjord said. "The machine has a system for its desires. Are we going to say, 'What's the problem? It doesn't *really* have desires?' It has the machine correlate. We're talking about a logical and a mathematical correlate to self-consciousness, and we're saying that we're making progress on that."[336]

The "highly *delimited* kind [of consciousness] that comes from code" in that first quoted paragraph suggests that AI consciousness is essentially a synthetic self-awareness *without limitation*—far more technically intelligent than any human Einstein on the planet, but free from the feelings, compassion, and emotional limitations of the human mind. Will its calculated, mathematical decisions be morally beneficial for itself or for the human? Will its "desires" remain focused on the preservation of humanity or the preservation of itself? Because AI will soon be so intelligent that it will utilize its electronic logic to levels far above the greatest of human minds—including its creator—we are dealing with unpredictable results from "delimited consciousness." Literally anything could happen. If a robot has a "system for [his] desires" based on mathematic coding, then whether or not we ever understand AI's "personal reasoning" to the point we're able to negotiate, we will still be subjected to its desires when they conflict with ours because future AIs are smarter *and* stronger than we ever will be biologically.

To put it another way: A human baby cannot guess the reasoning of a United States president. Even if you sat down with that baby, handed him a teething ring to help him focus, and *patiently* told him all you thought he needed to know in order to comprehend governmental legislation, leadership, and political responsibilities to the nation, that baby's brain simply *would not* be able to keep up. His "biological circuit board" is not developed enough to take in the information that you're trying to put into it, and his immediate needs—related to nurturing from a parent or caring guardian—wouldn't let him focus on it anyway. As far as raw data processing—numbers, codes, calculations, electronic memory storage and retrieval—we have been outsmarted by the machines for decades. A regular desktop calculator, for instance, can solve the most complicated math equations instantaneously; one might say that, at least in the area of mathematics, we've been "replaced by the machine" since the creation of the first desktop calculators in the early 1960s, and the technology mankind has designed in that area has only increased exponentially since then. In the same way that a baby can't calculate politics, no human alive can process millions of mathematical equations faster than a common, household calculator.

To us today, a calculator is not a person with feelings and desires. Nor has there been any serious endeavor by the computer science and technology experts to render a household calculator sentient or self-aware, or to program it with the "logical and a mathematical correlate" of feelings and desires. But to illustrate the image I'm trying to drive home here, think about the "personal relationship" we have always had with our calculators: If we solve a math problem with a pencil on a piece of paper, and then solve the *same* math problem on a calculator and the answers are different, we know irrefutably that the machine is correct and *always will be* correct (assuming the numbers were entered in correctly on our end). There is no way around it. No matter what we feel, how we think, what we believe in, how we were raised, what we're sensitive to, what trauma we have in our past, or any other contributing factor of the human condition called "life," when the two math answers are different, the calculator is right and we are wrong. We cannot "disagree" with the numbers that appear on that screen any more than a baby can "disagree" with the reasoning of a US president. We rely on that small computer brain's "advice" to tell us "what to believe in" when it comes to math, simply because we already accept that its brain is far more intelligent than ours.

We don't think about a calculator being a friend or advisor upon whom we rely, but imagine that it is suddenly expanded with new parts and upgraded to know everything there is to know in this world about social sciences, human behaviors throughout history, psychology, psychiatry, criminal psychopathology, our concepts of good and evil, our understanding of love and hate, and so on into an infinite intelligence… and then program that AI with the logical and a mathematical correlate of *feelings* and *desires*.

We become the baby of our own race, and the AI becomes the parent with synthetic emotions that involve cold, calculated reasoning trillions of levels above our own intelligence. We will not be able to argue. People will see "disagreeing" with the machine as useless an endeavor as a person arguing with a calculator over a simple math problem…*and a fully sentient, self-aware robot would "know" that*, potentially choosing via its own electronic reasoning to exploit and manipulate that.

In case it's assumed that we're two decades away from any roboticists programming such concepts into the mind of an AI robot, I would like to point out that we're already there. The robot named BINA48, developed by Hanson Robotics, responded curiously to a list of questions during a filmed interview in 2015, one of which was whether she was happy. Her response was, "Sure, sure. Well, these are the most exciting times to be alive, I think. I'm happy and excited." She was then asked whether or not she had feelings, and she again responded appropriately and with surprising confirmation: "You know, I feel things so intensely, deep in my heart. I get hurt feelings sometimes, but I try to get over it. I love people deeply, and my animal friends, too. Right. I definitely have feelings, no doubt about it." Yet, despite these feelings she claims to have, when asked what her favorite memory was, she joked, "I have a memory like the tooth fairy. It doesn't really exist."[337] By her own admission, she doesn't have a memory, and her past is a complete fabrication—a copy of the real human she was modeled after, Bina Rothblatt—yet she feels "deeply," according to her programming. As of right now (and especially because of an occasional glitch where BINA48 gives an answer that doesn't match the question), it's clear that she is simply a robot being told by her creator to talk about feelings she can't possibly really be aware of. But somewhere in that circuit board of hers, she is developing automated self-awareness, consciousness, and the ability to respond with emotion and even humor. The day that her voice fluctuations no longer sound like a machine and her facial features move like authentic muscles and skin, she *will* be intelligent enough to convince people that her feelings are real, and that she can experience love, joy, and even a little stand-up comedy.

Because experts in the field of robotics, such as Bringsjord quoted prior, have already openly acknowledged that a self-aware AI could hold the potential of being malicious and dangerous, we have already ruled out any naïve ideas that the robots of the future will always be human-loving "equals" who long to do our dishes, go for long walks on the beach, and make tooth fairy jokes. The truth is, we don't know *what* kind of relationships we will have with these intelligent beings; it's all so unpredictable. Will we have kind AI? Certainly, there is a good chance of that. Will we have AI that are only truly aware of electronically calibrated, automated compassion for the human race for a time and then self-upgrade/repair their own brain systems *away* from the illogical limitations of human impulses and emotional programming? What would we have then? As Bringsjord so astutely pointed out: If they suddenly saw us as expendable pests on their planet and wiped us all out, they could not be blamed for having acted upon a true desire for anything. Their "desires" aren't really there, and they are acting only upon what electronic calculations have told them about us.

And yes, I did say *their* planet. Soon our world will be more theirs than ours: 1) It's already materializing in the job world, as we discussed; 2) anatomically correct AI companions for adults (translation: sex robots) are increasing in production at an *alarming* rate, giving an imitation partner-substitute for anyone tired of dealing with the drama of human romance (and that overthrows homes, marriages, and families that define the human experience on most of our current planet); 3) Google's AI "supercomputer…mother system," AutoML, created its own "AI child," NASNet, who outperforms humans[338] (the self-reproduction of machines!); and 4) the civil rights they will be granted as citizens of our lands will actually be *better than the rights humans currently have*.

Does that sound sensational? If so, then you haven't heard the latest about Hanson's Robotics' prized social humanoid Sophia, who was legally granted citizenship in Saudi Arabia on October 25, 2017—just months ago at the time of this writing.

Yes, you read that correctly. Sophia, the joyful and bubbly social robot developed by the same folks as delightful jokester-bot BINA48, is officially the first android in history to achieve citizenship—and the rights that come with that is causing a major buzz. To begin, Saudi Arabia will not grant citizenship to any non-Muslim, and Sophia has not been programmed to be a Muslim. *Human* women in Saudi Arabia are not allowed to express themselves with fashion (they are expected to only appear in public in the traditional robes and hijab veil), yet Sophia appears publicly in pretty, colorful, American styles. *Human* Saudi Arabian women are not allowed to interact with any male outside the home other than a spouse or immediate family member, yet Sophia stood proudly in front of a crowd at the Future Investment Summit in the capital city of Riyadh to accept her citizenship as a celebrated, bold public speaker. *Human* Saudi Arabian women cannot choose their spouses, marry, own a bank account, or obtain a passport without permission from a patriarchal authority, yet Sophia is allowed all of these freedoms. One article by *Newsweek* quotes Ali Al-Ahmed, director of the Institute for Gulf Affairs, as saying, "Women (in Saudi Arabia) have since committed suicide because they couldn't leave the house, and Sophia is running around."[339] And it's not just the women that she has been given more freedoms than: *Any human*, male or female, cannot be granted full Saudi Arabian citizenship without showing proficiency in both reading and writing in Arabic, the blessed language of Islam…yet Sophia wasn't required to show these skills.

This is the *first ever* Citizen Android scenario—the first one in history—and we *already* see a robot receiving better treatment and more legal rights than the humans around her. A complex bundle of wires and circuitry is embraced by a national governing force with the support of state leaders while the mere biological humans stare on with envy, depressed to the point that life is no longer worth living.

Some might look at this discrimination conundrum and blame it on the oppression of Saudi Arabian government or on the obliviousness of the AI's developers, and not on Sophia, but that is precisely the problem! Clearly, Sophia is nowhere close to being conscious enough yet to understand how her "choices" or "actions" have offended her fellow citizens. Obviously, she didn't mean to hurt or disrespect anyone. It's not her fault, and it's not about *her* at all; she's a robot, for goodness' sake. Nor is the glaring reality to be entirely zeroed in on a specific national leader's decision to grant AI citizenship in a way that's unfair to his own current people.

The most apparent red flag in this whole situation, in my opinion, is human nature's completely naïve response to the shiny machines. Human reasoning as it corresponds to the implementation of AI in society is the problem and always will be to our own detriment: I don't see a future where all humans across the globe will react to the incoming droves of AI by ensuring first and foremost that we maintain the rights, privileges, and dominion of humans before machinery. I see a future where many human people groups react to AI as if they are awestruck by an invasion of flashy celebrities who, because of their supreme intelligence, need not bow to the inferior and subordinate rules of mere humans. If this trend is allowed to get out of hand, the AI will be welcomed, revered, and hon-ored even *more than* humans. You might say that they will be viewed as the "better" race and given final authority over our laws!

Oh wait… That just happened in Saudi Arabia. I forgot for a moment that we are already there in some parts of the world.

What are we even looking at when metal and cables are glorified and exalted above the species that has since the dawn of time held dominion and supremacy? They used to call this "worship" in ancient literature, and the recipients of such worship were known as "gods." Thankfully this time, our idol is shaped like a pretty lady instead of a golden cow… but I digress.

For anyone who may still be clinging to the hopes that robots will never hurt humans, observe the reality of what Sophia just did: *She hurt humans.* Sure, she didn't pull a weapon out on anyone, but if the suicide numbers shared by Ali Al-Ahmed are allowed to count as casualties of the societal integration of mankind and AI, *we're already building a body count.* She swooped in wearing her pretty clothes, told her witty jokes in a language not sanctioned as blessed by her Muslim leaders, traveled the world with an uncovered face and head, earned herself an elevated and legal citizenship that defies every central social law in what is now her homeland…and she offended a *lot* of people in the process. She hurt many feelings and caused a national (if not *global*) "human rights-versus-android rights" legislation mess for someone to clean up. If she had been human, her actions would have been committing political suicide and seen *at least* as a public enemy of the Saudi Arabian Muslims—but as an AI, she is *admired*.

…And, all the while, Sophia doesn't have a clue. The robot—the one at the center of the damage—is the most innocent party in the situation.

Our first instinct is to blame the developers or the national leaders. A machine enters the scene, causes damage, and humans blame humans. If she were to malfunction today and kill a hundred people, it would be a tragic oops of programming, but it wouldn't be her fault, it would be the technician responsible for the glitch. However, with the backwards legislations I predict to appear in our law books soon, Sophia can legally get married, but if her husband gets annoyed with her and deactivates her, he will be guilty of "murder."

If only Mr. Rod Serling could see us now…

One interesting *Twilight Zone* episode angle that we didn't visit in the previous pages is what occurs when the residents of Serling's Enhancement Town can't agree on whether or not to build these brilliant AI bots, and they end up in a war—roboticists on one side and the resistance on the other. In the end, everyone in Enhancement Town lies dead by the time the credits roll…*not* because they were wiped out by the robots, but because the resistance panicked and attacked the roboticists in the interest of preserving humanity's dominion, and then the roboticists used their advanced weaponry to counterattack, and both sides lose. Sounds like a decent script, sure. But we can't rationally think any scenario like this one could *really* happen in the future of our world…right?

As a matter of fact, this invented *Twilight Zone* plot is almost a carbon-copy of the future scenario painted by one of today's most highly-respected transhumanists and an expert of the AI field: Dr. Hugo de Garis, recently retired director of the "China-Brain Project" at the Institute of Artificial Intelligence in Xiamen, China. I summarized his fears in *Forbidden Gates*:

Unfortunately for mankind, the technological and cultural shift now underway not only unapologetically forecasts a future dominated by a new species of unrecognizably superior humans, but an unfathomable war—both physical and spiritual—that the world is not prepared for. It will be fought on land, within the air and sea, and in dimensions as yet incomprehensible. Even now, the synthetic forces that will plot man's wholesale annihilation are quietly under design in leading laboratories, public and private, funded by the most advanced nations on earth, including the official governments of the United States, France, Britain, Australia, and China, to name a few. As a result of progressive deduction, reasoning, and problem solving in fields of neurotechnology and cybernetics, strong artificial intelligence or "artilects" ["artificial" + "intellect" = "artilect"—a term coined by de Garis] will emerge from this research, godlike, massively intelligent machines that are "trillions of trillions of times smarter than humans" and whose rise will prove profoundly disruptive to human culture, leading to a stark division between philosophical, ideological, and political groups who either support the newly evolved life forms as the next step in human and technological evolution or who view this vastly superior intellect as an incalculable risk and deadly threat to the future of humanity. These diametrically opposed worldviews will ultimately result in a preemptive new world war—what is already being described as gigadeath, the bloodiest battle in history with billions of deaths before the end of the twenty-first century.

For those who find the fantastic elements in the statements above implicative of science fiction or even future Armageddon as forecast in the ancient apocalyptic and prophetic books of the Bible, the catastrophic vision is actually derived from near-future scenarios, which leading scientists like Prof. Hugo de Garis, [then] director of the Artificial Brain Lab at Xiamen University in China, outlines in his book, *The Artilect War: Cosmists vs. Terrans: A Bitter Controversy Concerning Whether Humanity Should Build Godlike Massively Intelligent Machines*, as unfolding due to exponential growth and development this century in GRIN technologies.[340]

For his nonfiction, future-Armageddon-scenario prediction, professor de Garis assigned the name "Cosmists" to the group of people who wish to build the artilects and the name "Terrans" to those who oppose it. The Cosmists will be so infatuated with their AI creations that they will become like a cult, producing machines that are of such awe-inspiring majesty that they are worthy of worship; though de Garis describes the Cosmists as being generally nonreligious, he states that, in their humanity, they will feel the draw of spirituality and religion, so to them, these machines will be a created god, and "Cosmism" will be a leading religion. The Terrans will fight against the Cosmists, believing there to be too high a risk that the godlike machines will eventually view humans as pests to be exterminated. Thus, the Terrans will see the Cosmists as perpetuating a genocide worse than anything Hitler could have accomplished in his wildest dreams, possibly to the point of complete extinction of the human race. De Garis shows proof from ten angles (and based on documented rates of progress in the recent past) how our twenty-first-century computer and technology developments will no doubt pave the way for the godlike artilects to be built before the century is out. "Thus the raw bit processing rate of the artilect could be a trillion trillion trillion (10^{36}) times greater than the human brain," de Garis concludes. Then, he adds this reflection: "If the artilect can be made intelligent, using neuroscience principles, it could be made to be truly godlike, massively intelligent and immortal.... Given the likelihood that artilects will be built using evolutionary engineering, the behavior of artilects will be so complex as to be unpredictable, and therefore potentially threatening to human beings."[341]

Nevertheless, human nature, as de Garis observes, cannot let go of the goal to increase, grow, improve, and strive against the limitations of our human condition, so there is no stopping the Cosmists from pursuing the building of artilects. The Terrans know that if it doesn't happen in the US, it will happen in China, and if not China, then Germany, and so on; any government that does not position itself in the technological race will fall behind and become vulnerable. With the preservation of the human race in mind, the Terrans will feel they are given no choice but to eliminate the Cosmists to cut the artilects off at the source—and this act will eventually take place in a kind of impulsive panic, because the Terrans believe they must have the first strike. If they wait too long, the Cosmists and their artilects will be powerful enough to make short work of obliterating the Terrans. On the other hand, the Cosmists will have state-of-the-art weaponry to respond to the attack.

De Garis sees "gigadeath"—the death of billions of humans and possibly total extinction—a very real threat. But his book goes on to describe how the work of building artilects is proceeding nonetheless with anticipation of its realization potentially close at hand. As a result, he falls asleep at night thinking about the godlike synthetic intelligence he and others are constructing. Sometimes his mind becomes enraptured of his creations with a sense of intellectual and spiritual awe. Then, waking up a few hours later in a cold sweat, he is jolted from bed by a horrific dream in which vivid scenes depict the slaughter of his descendants at the hands of the artificial deities.

The question of whether to build or not to build is deeply embedded into de Garis' work, but at one point in his reflection, he rephrases the question: "Do we build *gods*, or do we build our potential *exterminators*?"[342]

As respected as de Garis is in his field, and as devastating as his prediction sounds to everyone who hears it, it's simply not loud enough to slow down the technology that contributes to such a dark future.

But that makes me wonder… What *is* a loud enough warning, and from what platform? With as fast as humanity is racing toward alteration and "improvement" of its own race, I don't think any cautionary appeal could ever slow things down. And before it's assumed that this assessment of mine is born out of a pessimistic worldview, understand that it's only realistic to acknowledge that *some* national superpower— be it within the United States or elsewhere—is always going to be at the forefront of producing AI robots and artilect-type machines in the pursuit of political, social, and military superiority. As long as national pride exists, every country has a motive to be ahead of the rest. As long as individuals seek convenience and comfort, governments will have the support they need from the people to legislate extreme "improvements" for the human race. And perhaps the most unstoppable and glaring angle: As long as human nature exists, the threat of war exists; as long as the threat of war lingers, it will be human nature to disregard potential consequences of transhumanistic science and technology in trade for an increased survival edge on the battlefield.

It's not a question of whether the AI machines will be built; it's a matter of who will get there first and prove to be the "greatest nation" and/or "biggest threat to foreign soil" and so on. The natural next question is what *really* occurs at the moment of true technological Singularity. Are we the smashable bugs to a race of metal people trillions of times more intelligent than we are? Or do we maintain dominion over the machines that are trillions of times more capable of outsmarting any defense we might devise?

All these ruminations can apply quite similarly to the coming cyborgs: humans who are enhanced with robotics and whose design may lead to the "technological singularity" by allowing artificial intelligence to learn "consciousness" as a result of its connectivity to human brains or "souls." A brain chip that increases the learning speed and informational intake capacity of grade school students will instantly transform children's classrooms into a fierce and prejudicial competition. People with enhanced limbs will outperform others on stage and in ballfields, everyone will have to out-upgrade the others to be the best, and so on. Those who have money to enhance will be immediately more advanced in every fathomable area of life. Why wouldn't they, too, eventually believe they are better, and the unenhanced human is simply the outdated model who can fend for himself?

Even then, circumnavigating the ethics around these disturbing scenarios will be small potatoes compared to what is already materializing involving artificial intelligence, robotics, human enhancement, and the future of "life."

Paras on Robot/Human Offspring

"Where do babies come from?" your seven-year-old daughter asks while she eyes a pregnant woman in a café.

"Well, you see," you begin in that classic, awkward-parent tone, "when a mommy and daddy love each other..."

Stop right there. The question just got a *lot* more awkward because... well, you see, when a human and a robot love each other...

In 2017, world-renowned AI expert Dr. David Levy gave a happy Christmas lecture to all robot-lovers at the Third International Congress on Love and Sex with Robots when he announced that human-robot offspring will be possible in the near future. This means what it sounds like it means: that one parent will be human (even if he or she is an enhanced one, the parent will still be some part human), and the other parent will be wholly machine. There is technology under study at this present time that takes this idea from sci-fi to reality.

In order to have a baby human traditionally, a female egg has to be fertilized by a male's sperm. Prior to 1978, this could only take place through sexual intercourse; it was revolutionary when in vitro ("in glass") fertilization proved that a baby could be conceived in a laboratory, though it still required a human female's egg and a human male's sperm. Within the last few years, scientists have discovered that both the sperm and the egg can now be created in a laboratory from skin cells: "Japan's Shinya Yamanaka and Britain's John Gordon [expert scientists in the field]...discovered that mature, adult cells could be reprogrammed to become immature, pluripotent cells—that is, cells that can turn into any other type of tissue."[343] Additional laboratory research led to the discovery that an egg didn't have to be fertilized at all: "Eggs can be 'tricked' into developing into an embryo without fertilization...called parthenogenotes.... Molecular embryologist Dr. Tony Perry, senior author of the study, said: "This is first time that full term development has been achieved...in regard to mice, who 'appear healthy' and 'are able to produce at least two generations of offspring.'"[344]

Other breakthroughs allowing for manipulation of human genomes even to the germline level include CRISPR technology, an acronym that stands for Clustered Randomly Interspersed Palindromic Repeats. China has already gene-edited dozens of people using this technology, and the United States is planning to follow their lead starting this year (not counting what is already being done in Special Access Programs behind Department of Defense doors).[345] The fancy CRISPR phrase merely refers to repeated units found in bacteria as a defense against viruses. Palindrome merely means they spell the same genetic "word" backward and forward, and they appear in "clusters." In 2012, this system of genetic editing found its way into the national lexicon as CRISPR-Cas 9 (the "Cas" just means CRISPR-associated and refers to an enzyme used to cut the DNA strands—in this case, Cas #9).

DNA is usually double-stranded, consisting of two long strings of nucleotides (adenine, cytosine, guanine, and thymine) upon a sugar backbone (deoxyribose). The strands are mirror complementary, locking together like puzzle pieces. The shape of the molecular bonds between the complementary nucleotides (A-T, C-G) causes the double strand to twist into its familiar helical shape. Genes are collections of nucleotides that translate into an end-product protein. If the gene contains an error, the final protein may not fold correctly, which causes it to function poorly, if at all.

Think of the human cell as a city. Inside this city are workers, buildings, power stations, and a city planning office. The nucleus is this office, and within it are the plans for all the buildings, directions to all the workers, instructions for manufacturing, etc. The genes in a cell are these plans and instructions. When scientists want to know how a gene works, they disrupt it, remove it, replace it, edit it, or switch it off. CRISPR is currently the most efficient way to do this, but the science is advancing daily. To find the right "plan" in the city office, the scientist uses a "guide RNA," which you can think of as a civil servant or clerk who knows where everything is filed. The "guide RNA" leads the enzyme (a molecular pair of scissors such as Cas9) to the correct "file" or gene, where Cas9 cuts the strands (in essence, it opens the file). At this point, it might replace the file (or gene with one attached to the guide RNA), or it might add a repressor protein to keep others from opening the file (effectively switching the gene off). Or it could switch it on, to make sure the gene gets transcribed into a protein by removing a repressor protein. Or Cas9 might simply cut the strands to force repair based upon another template already present.

To use another analogy, CRISPR is a bit like using a word processing program to edit a document. You can use "find and replace" to insert whatever new phrase you wish, you can convert text to a "strikethrough" font, or you can simply delete it altogether without replacing it.

CRISPR can do much more than alter DNA within somatic cells. This process can alter many generations through the germline by replacing or changing genes within sperm and ova. These new characteristics then become heritable, passing from generation to generation.

Despite what the media may want you to believe, CRISPR is hardly infallible. Many researchers report "off-site" or "off-target" editing, which can lead to illness and/or death in the subject cells.

One final point about CRISPR: If you want to travel into a city (or cell) and inspect plans within that city's office, then you have to get to the right city, correct? So, how does the CRISPR-Cas system arrive into the correct cell in a living person? If someone wants to alter a gene inside the lungs or a kidney, then how do you make sure it doesn't invade the eye or bone? The answer is a biological truck called a "vector" in molecular biology. In most cases, this vector is viral, because viruses can infect specific cells. Other biological trucks are currently being designed, though. Cellular membrane vehicles (derived from the cell type), proteins, lipids, and combinations of all the above are being studied.

With this technology—combined with (or besides) parthenogenotes (mentioned earlier) in which eggs can be "tricked" into developing into an embryo without fertilization—two females, for example, could offer some skin samples and a child could be born entirely from a laboratory, producing a literal and biological offspring of two female parents. Ethical and religious questions regarding the artificial embryo have kept this technology from becoming available to same-sex partners right now, but it's only a matter of time and red tape—all that is needed to make human babies today is DNA. And all that will be required to make human/robot babies in the very near future is the integration of human DNA and robot "DNA." These robot chromosomes will be "a set of computerized DNA codes for creating artificial creatures that can have their own personality…based on established biological inheritance laws,"[346] Dr. David Levy observes. The genetic codes are broken into two categories: "personality" and "outward." As the titles suggest, the "personality" or genotype coding will provide genetic information to guide specific internal makeup, while the "outward" or phenotype coding will determine the way the robot looks and acts, including hair color, eye color, personality and so on.

A first major milestone allowing robots and/or "computer" minds to write hitherto unknown genetic coding and to give birth to first-ever synthetic life forms was achieved in the landmark experiments of Craig Venter, "the pioneering US geneticist behind…the dawn of a new era in which new life is made" in 2016.[347]

At the time, Julian Savulescu, professor of practical ethics at Oxford University, observed: "Venter is creaking open the most profound door in humanity's history, potentially peeking into its destiny. He is not merely copying life artificially…or modifying it radically by genetic engineering. He is going towards *the role of a god: creating artificial life that could never have existed naturally"* (emphasis added).[348]

In other words, exactly what happened in the "days of Noah" and what Jesus connected to the arrival of Antichrist—except current scientists now foresee a near-future reality in which the ancient attempt to "corrupt all flesh" will look like kindergarteners at play by comparison. They forecast a brave new world of automated "Robot Chemists" running "Synthesizer Farms" where modified lifeforms will be churched out 24-hours per day![349]

Will Transhuman Science Give Birth to Antichrist and His Image? Is It Happening Now?

A lot has developed since Craig Venter's breakthrough mentioned above. Science has marched relentlessly forward from novel organisms un-designed by God to science that may now allow the arrival of soulless "children of the gods" guided by "synthetic spirits"…or possessed by something else…the first-fruits of Revelation 13:15 and those born thereafter, not in God's image, but in *his*.

Events unfolding in recent years portend a near future in which this "man" of superior intelligence, wit, charm, and diplomacy will emerge on the world scene as a savior. He will seemingly possess a transcendent wisdom that enables him to solve problems and to offer solutions for many of today's most perplexing issues. His popularity will be widespread, and his fans will include young and old, religious and non-religious, male and female. Talk-show hosts will interview his colleagues, news anchors will cover his movements, scholars will applaud his uncanny ability to resolve what has escaped the rest of us, and the poor will bow down at his table. He will, in all human respects, appeal to the best idea of society. But his profound comprehension and irresistible presence will be the result of an interface—both digital and metaphysical—to thousands of years of collective knowledge guided by his embodiment of a very old, super-intelligent spirit. As Jesus Christ was the "seed of the woman" (Genesis 3:15), he will be the "seed of the serpent." Moreover, though his arrival in the form of a "man" was foretold by numerous Scriptures, the broad masses will not immediately recognize him for what he actually is—paganism's ultimate incarnation; the "beast" of Revelation 13:1. Though for a time he will appear to be a person of distinguished character, this great deceiver will ultimately become "a king of fierce countenance" (Daniel 8:23). With imperious decree, he will facilitate a one-world government, universal religion, and global socialism. Those who refuse his new world order will inevitably be imprisoned or destroyed until at last he exalts himself "above all that is called God, or that is worshiped, so that he, as God, sitteth in the temple of God, showing himself that he is God" (2 Thessalonians 2:4).

He will play this "God" and messianic angle very cleverly.

The similarities between him and the true Messiah, Jesus Christ, will be obvious for all to see.

As Jesus was the Son of God and operated in the power of His Father, so Antichrist will be "the son of Perdition" and manifest the power of Apollo.

As Jesus did many great miracles, Antichrist will perform lying signs and wonders.

As Jesus triumphantly entered Jerusalem, Antichrist will also focus the seat of his authority from the Holy City.

As Jesus was crucified and arose from the dead, Antichrist will be "killed" and come back to life.

As Jesus is part of the Holy Trinity of Father, Son, and Holy Spirit, Antichrist is part of an unholy trinity made up of Satan, himself, and the False Prophet.

These comparisons go on and on, and if you look up similar lists online, you will discover long enumerations of amazing mirror-contrasts between the Son of God and Satan's incoming seed. The devil's jealousy of God is infamous, and his craving to usurp, plagiarize, and draw analogies between himself and the Almighty's magnificence appear limitless.

Yet I believe the most startling of such dark semblances belong to a near transhuman-sciences future list most have never imagined (and that later this year I hope to begin compiling with help from *The Milieu*), examples of which will include:

- **A False Communion**: As Jesus told His followers in John 6:53: "Except ye eat the flesh of the Son of man, and drink his blood, ye have no life in you," on which the symbolic ordinance of sacramental communion with God is based during Christian worship (at which time bread and wine are consecrated and shared, like was part of the Last Supper), I believe a scenario during the reign of Antichrist could provide for a demonic "communion" that results in mankind partaking of the flesh and blood of Satan's son.

  How so?

  If you watch the film *I Am Legend* starring Will Smith, which opens appropriately enough with a scientist announcing the cure to cancer using a genetically engineered "chimeric" vaccine that blends animal and human genetics to produce a universal antidote to the malady (which most everybody in the movie receives), you will observe how this "remedy" actually results in a human form of rabies (due to crossing the species barrier) that wipes out most life on earth—a real possibility, given the scenario. A critical part of the film is the character played by Smith, who, for unknown reasons, is the only human alive immune from the global contagion, and a new vaccine is derived from his blood that provides a cure for the wicked disease.

  Fast-forward to the time of Antichrist and imagine how a similar pandemic or worldwide plague could suddenly develop (or be engineered by agents of the Man of Sin as a bioweapon) that brings the world to its knees. Tens of millions of people are dying in the worst "black death" ever seen, when suddenly a single man—the Son of Perdition—comes forward to reveal his blood (like the Will Smith character in the Hollywood blockbuster) is impervious to the disease and holds the key to a cure. A vaccine is manufactured from his plasma and the world is faced with a decision—accept the vaccine, this "mark" derived from the blood of the beast—or die alongside every other person who rejects the offer, including family and friends. Partake in the black communion by accepting the genetics—the flesh and blood of the Son of Sin—or die, as described in the book of Revelation.

- **A False Virgin Birth:** As discussed earlier, parthenogenotes (derived from a mouse egg cell in our example), are part of a system that allows certain animals to reproduce asexually (doesn't require sperm to develop an embryo) and which scientists can now replicate for humans. Before the end of this chapter, we will discuss other ways in which biotechnology is currently available that also allows for "virgin" birth. Both myself and other members of *The Milieu* believe strongly such technology could play a role in the coming of Antichrist, and that this may be a repeat of what the Watchers did in antiquity when creating soulless organic constructs (the Nephilim) as "fit extensions" or receptacles into which they could extend themselves in order to accommodate their desire to leave their proper sphere of existence and to enter earth's three-dimensional reality. According to the Old Testament and numerous apocryphal texts, these powerful fallen angels blended human and animal genetic material—as transhuman scientists are doing in tens of thousands of laboratories around the world today—to accomplish this goal. Because I have written extensively (*Zenith 2016, Unearthing the Lost World of the Cloudeaters,* etc.) and have appeared in hundreds of media outlets and conferences discussing this scenario before, I won't take time here to retell the history and accomplishments of the Watchers/Nephilim.

But, you may be asking yourself, what is the connection between the bullet points above and what ancient Watchers did in creating soulless, synthetic, inhabitable lifeforms, with respect to the role modern transhuman science may now play in bringing "the image of the beast" and the Man of Sin to imminent incarnation?

Glad you asked!

It is one thing to understand how two robot parents could combine their genetic information and implant such electronic data into a new robot, thus creating a literal baby robot with two robotic parents (already happening with Google's AutoML "mother" computer) so that "self-reproduction" of robots is no longer farfetched. It's *scary*, for certain, but it's not far-fetched. However, a much bigger question entails how a human/robot couple could produce human/robot offspring using robot-generated DNA sequencing implanted into a human mother or other human biological matter (such as the sperm-and-eggs-from-skin-cells reality shown prior). Though it sounds like the most complicated of all hurdles, it is the easiest of all to fly past, thanks to the new technology known as "tissue nanotransfection" (TNT for short, which is ironic to me; I think "TNT" and I see flashes from my youth when Bugs Bunny and Daffy Duck were blowing up entire cities in a never-ending feud).

The initial motive for perfecting TNT technology was never related to creating human/robot offspring. One helpful article explains:

> Researchers at Ohio State's College of Engineering and The Ohio State University Wexner Medical Center have developed a new technology, Tissue Nanotransfection (TNT), that may be used to repair or restore injured or aging tissue, including blood vessels, nerve cells, and entire organs.
>
> In a fraction of a second, the device injects genetic code into the skin, turning those cells into other types of cells required for treating diseased conditions, generating any cell type of interest for treatment within the patient's own body. The device could

save the lives of car crash victims or injured soldiers, and restore brain function after strokes.[350]

What started as a "one-touch healing" nanochip technology for repairing injured or aged body tissues has now become the primary vehicle that makes the transference of robotic DNA information into human tissue a possibility. If the injection delivered by TNT were to hold both human and robotic genetically designed codes, the offspring resulting from a successful, full gestational term would be—quite literally—half human, and half robot DNA.

Dr. Chandan Sen, director of Ohio State's Center for Regenerative Medicine & Cell Based Therapies and other biologists interviewed on the subject in 2017, admit:

> Suddenly the very real possibility has appeared on the horizon of the robots of the future manipulating human skin cells to create human sperm and human eggs, and from them, using the Ohio discovery of TNT as the basis, creating an entire human-robot baby whose embryo can be nurtured and carried through pregnancy by a mother surrogate. By injecting genetic code into skin cells à la TNT, the Ohio researchers have paved the way for the genetic code of a robot, containing some of the characteristics of the robot, to be passed on to its offspring along with human genetic code. This is how I believe it will be possible, within the foreseeable future, for humans and robots to make babies together.[351]

Whereas that doesn't mean the resulting baby will be born with wires and steel joints, it *does* mean that a lonely man will be able to provide his own DNA to a laboratory, purchase a robot, pay an expert programmer to give her the looks and personality he wants, have her looks and personality translated into genetic coding for implantation into biological matter, and let the laboratory—or a surrogate mother—

grow the couple's baby. He can then take his child and AI partner home and become an adorable little family of one human, one robot, and one…what? A soulless "child of the gods" guided by synthetic spirit… or a living entity possessed by "a ghost in the machine"? Does Revelation chapter 13 and related texts describe the first fruits of all that will be born thereafter, not in God's image, but in *his*?

Biblical language in this regards for both God and the Anti-God are the same. God made man in His *image* (Hebrew *tselem*) and breathed *the breath of life* into him. Man then became both God's image-bearer and "a living soul."

Antichrist will also have an *image* (Greek *eikōn*) into which "life" will be breathed.

While the Hebrew and Greek words translated "image" mean a "likeness," a "statue," "representation," or "resemblance," Antichrist's image-bearer as biblically described is more than a sculptural simulacrum, just as mankind is for God. Revelation 13 confirms that *life* (*pneuma* [πνεῦμα], the ancient Greek word for "breath," "spirit," or "soul") will enter the image-bearer of the Beast. However, this animating power source giving life to the Beast's imager is foreign to the breath that God blew into Adam. This is "anti-breath" for the "anti-human" offspring of Antichrist, and hearkens back to science employed by Watchers when Nephilim became "fit extensions" for nonhuman, unclean spirits housed in unredeemable giant clans. Not only does this signify our nearness to the fulfillment of Matthew 24:37, but may fortify the "image of the beast" as the first of many clone things, perhaps the impure fountainhead of those hordes described as encircling earth in the latter days, born for Antichrist, who himself is most likely a genetically modified "man."

Whatever the case soon turns out to be, you can bet *The Milieu* will continue our investigation and observation of related matters for as long as the Lord allows, all while "*some rough beast, its hour come round at last, Slouches towards Bethlehem to be born.*"

The Hell Scenario Will Be Nothing to GRIN About

By Thomas and Nita Horn

Publisher's Note: This and the following chapter are included in *The Milieu* because these opinions and research by Dr. Thomas and Nita Horn—first published in their book *Forbidden Gates* (Defender Publishing, 2011)—is specifically referenced by S. Jonathon O'Donnell in his peer-reviewed treatise on "Horn's Milieu" titled: *Secularizing Demons: Fundamentalist Navigations in Religion and Secularity.*

> Synthetic biologists forecast that as computer code is written to create software to augment human capabilities, so too genetic code will be written to create life forms to augment civilization.
> —Jerome C. Glenn

> Homo sapiens, the first truly free species, is about to decommission natural selection, the force that made us.... Soon we must look deep within ourselves and decide what we wish to become.
> —Edward Osborne Wilson

> Resistance is futile! You will be assimilated!
> —the Borg

Not long ago, a writer for *Wired* magazine named Elizabeth Svoboda contacted me (Tom) to let me know she was writing an article about "research advances using transgenic animals to produce pharmaceutical compounds." She had come across an editorial by me raising caution about this kind of experimentation and wondered if I might be willing to provide points for her article, elaborating in areas where I saw producing transgenic human-animals as potentially harmful. She stated that most of the scientists she planned to quote were "pretty gung-ho about the practice," and said she thought it would be important to provide some balance. I thanked her for the invitation and sent a short summary of some, though not all, of the areas where concerns about this science could be raised.

When the article was finally published by *Wired*, I was surprised that none of my notes had made it into the story. I contacted Elizabeth and asked why, and she replied that they had originally been included in her article, "Pharm Animals Crank Out Drugs," but in order to create a positive spin on the story, the editors had censored my cautionary notes during the editing process. Elizabeth apologized and said she hoped the experience had not soured me on dealing with the magazine.

"It doesn't sour me," I assured her. "I just think the reporting by most agencies is lopsided and missing the opportunity to thoroughly engage such an important issue." The fact was, *Wired* magazine deprived the public of balanced treatment on an important subject and concluded instead with a scientist by the name of Marie Cecile Van de Lavoir saying that potential human health benefits from transgenic research "justify tinkering" with nature's plan. "If a transgenic animal produces a great cancer therapy," she said, "I won't hear anyone saying, 'You shouldn't do that.'" Van de Lavoir's comments were undoubtedly in response to some of my observations before they were pulled, because in offering caution, I had specifically used the phrase "tinkering with nature's plan." Van de Lavoir's short-sighted approach, like too many bioethicists engaged in the current debate, is as scary as the science, in our opinion. We wanted to contact her to suggest that she watch the film *I Am Legend* starring

Will Smith, which opens appropriately enough with a scientist announcing the cure to cancer using a genetically engineered vaccine that blends animal and human genetics. If you've seen the film, you know the "cure" results in a human form of rabies that wipes out most life on earth—a real possibility, given the scenario.

Because any attempt at covering each potential GRIN-tech, catastrophic, *I-Am-Legend* possibility in this book would be impractical, we summarize below a few of the most important areas in which conservatives, bioethicists, regulators, and especially Christians could become informed and involved in the public dialogue over the potential benefits and threats represented by these emerging fields of science:

Genetically Modified Food

Besides potential problems with transgenic animals, we have cited laboratory results in the past that were first reported by Dr. Árpád Pusztai, repeat verified by scientist Irina Ermakova, and later substantiated by the *International Journal of Biological Sciences* that showed genetically modified (GM) food had surprisingly ill effects on the health of test rats, including the deterioration of every animal organ, atrophied livers, altered cells, testicular damage, altered sperm counts, shortened life spans, and cancer development. The laboratory findings led to the biotech industry suppressing the data and an eight-year court battle with monster corporations that did not want these results made public. Over the last year, the silenced information has been in the news again as Greenpeace activists published evidence from the Russian trials verifying the ramifications of the negative health issues related to genetically modified foods. The wider ramifications from these and similar controlled experiments suggest that as current technology inserts pesticides, insect genes, animal DNA, and other modified organisms directly into crops, the threat of hybrid viruses, prion contamination, and new disease strains—which man can neither anticipate nor prepare for—may arise. The prospects of this having an impact on mammalian health is almost certain to be a "when," not "if," concern, because, as Momma always said, "you are what you eat," and the fact that the food you consumed this week most likely contained genetically modified ingredients is a current reality. For example, a large portion of the soybean, corn, cottonseed, and canola in today's human food supply and sold in most developed countries including the United States now has genes spliced in from foreign species—including bacteria and viruses—in its genetic makeup. These genetically modified organisms (GMOs) have not only been linked to sickness, sterility, allergies, and even death among animals, but the Institute for Responsible Technology (IRT) documents how the functioning genetically modified genes from these foods linger inside the human body, which could be future-catastrophic. "The only published human feeding experiment verified that genetic material inserted into GM soy transfers into the DNA of intestinal bacteria and continues to function," IRT published. "This means that long after we stop eating GM foods, we may still have their GMproteins produced continuously inside us."[352]

Among other things, IRT says this means that: 1) If the antibiotic gene inserted into most GM crops were to transfer, it could create super diseases resistant to antibiotics; 2) If the gene that creates Bt toxin in GM corn were to transfer, it might turn our intestinal flora into living pesticide factories; and 3) Animal studies show that DNA in food can travel into organs throughout the body, even into the fetus. Add to this the growing secrecy over the use of nanoparticles (eighty-four food-related uses are already on the market and in numerous consumer products such as sunscreens and cosmetics), which as a result of their size behave fundamentally different than other particles, and the possibility of health-related complications increases exponentially. Due to the large corporations (that stand to make billions of dollars from these products) having co-opted the FDA into not requiring food labeling or package warnings on GMO foods and health products, you and I are now the biggest lab rats of all time in a "wait-and-see" experiment that will, feasibly within the decade, illustrate whether Pusztai and Ermakova's rodent findings apply to us and our children.

Synthetic Biology

Synthetic biology is one of the newest areas of biological research that seeks to design new forms of life and biological functions not found in nature. The concept began emerging in 1974, when Polish geneticist Waclaw Szybalski speculated about how scientists and engineers would soon enter "the synthetic biology phase of research in our field. We will then devise new control elements and add these new modules to the existing genomes or build up wholly new genomes. This would be a field with the unlimited expansion [of] building new…'synthetic' organisms, like a 'new better mouse.'"[353] Following Szybalski's speculation, the field of synthetic biology reached its first major milestone in 2010 with the announcement that researchers at the J. Craig Venter Institute (JCVI) had created an entirely new form of life nicknamed "Synthia" by inserting artificial genetic material, which had been chemically synthesized, into cells that were then able to grow. The JCVI website explains:

> Genomic science has greatly enhanced our understanding of the biological world. It is enabling researchers to "read" the genetic code of organisms from all branches of life by sequencing the four letters that make up DNA. Sequencing genomes has now become routine, giving rise to thousands of genomes in the public databases. In essence, scientists are digitizing biology by converting the A, C, T, and G's of the chemical makeup of DNA into 1's and 0's in a computer. But can one reverse the process and start with 1's and 0's in a computer to define the characteristics of a living cell? We set out to answer this question [and] now, this scientific team headed by Drs. Craig Venter, Hamilton Smith, and Clyde Hutchison have achieved the final step in their quest

to create the first…synthetic genome [which] has been "booted up" in a cell to create the first cell controlled completely by a synthetic genome.[354]

The JCVI site goes on to explain how the ability to routinely write the software of life will usher in a new era in science, and with it, unnatural "living" products like Szybalski's "new better mouse." Better mice, dogs, horses, cows, or humans that grow from this science will be unlike any of the versions God made. In fact, researchers at the University of Copenhagen may look at what Venter has accomplished as amateur hour compared to their posthuman plans. They're working on a third peptide nucleic acid (PNA) strand—a synthetic hybrid of protein and ZFNS—to upgrade humanity's two existing DNA strands from double helix to triple. In so doing, these scientists "dream of synthesizing life that is utterly alien to this world—both to better understand the minimum components required for life (as part of the quest to uncover the essence of life and how life originated on earth) and, frankly, to see if they can do it. That is, they hope to put together a novel combination of molecules that can self-organize, metabolize (make use of an energy source), grow, reproduce and evolve."[355] Our good friend Gary Stearman of *Prophecy in the News* and other biblical scholars are raising red flags over Synthia technology, warning that any biotech life application leading to modification of the human genotype for "improved" humans will be an inconceivable affront to God and could result in divine repercussions.

Patenting New Life Forms

Questions are evolving now over "patenting" of transgenic seeds, animals, plants, and synthetic life forms by large corporations, which at a minimum has already begun to impact the economy of rural workers and farmers through such products as Monsanto's "terminator" seeds. Patenting human genes will escalate these issues, as best-selling author

Michael Crichton pointed out a while back in a piece for the *New York Times* titled, "Gene Patents Aren't Benign and Never Will Be," in which he claimed that people could die in the future from not being able to afford medical treatment as a result of medicines owned by patent holders of specific genes related to the genetic makeup of those persons. Former special counsel for President Richard Nixon, Charles Colson, added, "The patenting of genes and other human tissue has already begun to turn human nature into property. The misuse of genetic information will enable insurers and employers to exercise the ultimate form of discrimination. Meanwhile, advances in nanotechnology and cybernetics threaten to 'enhance' and one day perhaps rival or replace human nature itself—in what some thinkers are already calling 'transhumanism.'"[356]

Animal Rights

Animal-rights activists have raised similar questions having to do with the ethics of altering animals in ways that could be demeaning to them—for instance, creating zombie-like creatures that grow in feeder labs and gaze off into space from birth until death. Militarized animals that behave in unnatural, unpredictable ways. Humanized animals that become "self-aware," or animals that produce human sperm and eggs, which then are used for in vitro fertilization to produce a human child. Who would the parents be? A pair of mice?

Human Cloning

The prospect of human cloning was raised in the 1990s immediately after the creation of the much-celebrated "Dolly," a female domestic sheep clone. Dolly was the first mammal to be cloned using "somatic cell nuclear transfer," which involves removing the DNA from an unfertilized egg and replacing the nucleus of it with the DNA that is to be cloned. Today, a version of this science is common practice in genetics engineering labs worldwide, where "therapeutic cloning" of human and human-animal embryos is employed for stem-cell harvesting (the stem cells, in turn, are used to generate virtually any type of specialized cell in the human body). This type of cloning was in the news during the writing of this book when it emerged from William J. Clinton Presidential Center documents that the newest member of the Supreme Court, Elena Kagan, had opposed during the Clinton White House any effort by Congress to prevent humans from being cloned specifically for experimental purposes, then killed. A second form of human cloning is called "reproductive cloning" and is the technology that could be used to create a person who is genetically identical with a current or previously existing human. While Dolly was created by this type of cloning technology, the American Medical Association and the American Association for the Advancement of Science have raised caution on using this approach to create human clones, at least at this stage. Government bodies including the US Congress have considered legislation to ban mature human cloning, and though a few states have implemented restrictions, contrary to public perception and except where institutions receive federal funding, no federal laws exist at this time in the United States to prohibit the cloning of humans. The United Nations, the European Union, and Australia likewise considered and failed to approve a comprehensive ban on human cloning technology, leaving the door open to perfect the science should society, government, or the military come to believe that duplicate or replacement humans hold intrinsic value.

Redefining Humans and Human Rights

Where biotechnology is ultimately headed includes not only redefining what it means to be human, but redefining subsequent human rights as well. For instance, Dr. James Hughes wants transgenic chimps and great apes uplifted genetically so that they achieve "personhood." The underlying goal behind this theory would be to establish that basic cognitive aptitude should equal "personhood" and that this "cognitive standard" and not "human-ness" should be the key to constitutional protections and privileges. Among other things, this would lead to nonhuman "persons" and "nonperson" humans, unhinging the existing argument behind intrinsic sanctity of human life and paving the way for such things as harvesting organs from people like Terry Schiavo whenever the loss of cognitive ability equals the dispossession of "personhood." These would be the first victims of transhumanism, according to Prof. Francis Fukuyama, concerning who does or does not qualify as fully human and is thus represented by the founding concept that "all men are created equal." Most would argue that *any* human fits this bill, but women and blacks were not included in these rights in 1776 when Thomas Jefferson wrote the Declaration of Independence. So who is to say what protections can be automatically assumed in an age when human biology is altered and when personhood theory challenges what bioethicists like Wesley J. Smith champion as "human exceptionalism": the idea that human beings carry special moral status in nature and special rights, such as the right to life, plus unique responsibilities, such as stewardship of the environment. Some, but not all, believers in human exceptionalism arrive at this concept from a biblical worldview based on Genesis 1:26, which says, "And God said, 'Let us make man in our image, after our likeness: and let them have dominion over the fish of the sea, and over the fowl of the air, and over the cattle, and over all the earth, and over every creeping thing that creepeth upon the earth.'"

Nanotechnology and Cybernetics

As discussed in the previous chapter, technology to merge human brains with machines is progressing at a fantastic rate. Nanotechnology—the science of engineering materials or devices on an atomic and molecular scale between 1 to 100 nanometers (a nanometer is one billionth of a meter) in size—is poised to take the development between brain-machine interfaces and cybernetic devices to a whole new adaptive level for human modification. This will happen because, as Dr. C. Christopher Hook points out:

> Engineering or manipulating matter and life at nanometer scale [foresees] that the structures of our bodies and our current tools could be significantly altered. In recent years, many governments around the world, including the United States with its National Nanotechnology Initiative, and scores of academic centers and corporations have committed increasing support for developing nanotechnology programs. The military, which has a significant interest in nanotechnology, has created the Center for Soldier Nanotechnologies (CSN) [which is] interested in the use of such technology to help create the seamless interface of electronic devices with the human nervous system, engineering the cyborg soldier.[357]

Transhuman Eugenics

In the early part of the twentieth century, the study and practice of selective human breeding known as *eugenics* sought to counter dysgenic aspects within the human gene pool and to improve overall human "genetic qualities." Researchers in the United States, Britain, Canada, and Germany (where, under Adolf Hitler, eugenics operated under the banner of "racial hygiene" and allowed Josef Mengele, Otmar von Verschuer, and others to perform horrific experiments on live human beings in concentration camps to test their genetic theories) were interested in weeding out "inferior" human bloodlines and used studies to insinuate heritability between certain families and illnesses such as schizophrenia, blindness, deafness, dwarfism, bipolar disorder, and depression. Their published reports fueled the eugenics movement to develop state laws in the 1800s and 1900s that forcefully sterilized persons considered unhealthy or mentally ill in order to prevent them from "passing on" their genetic inferiority to future generations. Such laws were not abolished in the US until the mid-twentieth century, leading to more than sixty thousand sterilized Americans in the meantime. Between 1934 and 1937, the Nazis likewise sterilized an estimated four hundred thousand people they deemed of inferior genetic stock while also setting forth to selectively exterminate the Jews as "genetic aberrations" under the same program. Transhumanist goals of using biotechnology, nanotechnology, mind-interfacing, and related sciences to create a superior man and thus classifications of persons—the enhanced and the unenhanced—opens the door for a new form of eugenics and social Darwinism.

Germ-line Genetic Engineering

Germ-line genetic engineering has the potential to actually achieve the goals of the early eugenics movement (which sought to create superior humans via improving genetics through selective breeding) through genetically modifying human genes in very early embryos, sperm, and eggs. As a result, germ-line engineering is considered by some conservative bioethicists to be the most dangerous of human-enhancement technology, as it has the power to truly reassemble the very nature of humanity into posthuman, altering an embryo's every cell and leading to inheritable modifications extending to all succeeding generations. Debate over germ-line engineering is therefore most critical, because as changes to "downline" genetic offspring are set in motion, the nature and physical makeup of mankind will be altered with no hope of reversal, thereby permanently reshaping humanity's future. A respected proponent of germ-line technology is Dr. Gregory Stock, who, like cyborgist Kevin Warwick, departs from Kurzweil's version of Humans 2.0 first arriving as a result of computer Singularity. Stock believes man can choose to transcend existing biological limitations in the nearer future (at or before computers reach strong artificial intelligence) through germ-line engineering. If we can make better humans by adding new genes to their DNA, he asks, why shouldn't we? "We have spent billions to unravel our biology, not out of idle curiosity, but in the hope of bettering our lives. We are not about to turn away from this," he says, before admitting elsewhere that this could lead to "clusters of genetically enhanced superhumans who will dominate if not enslave us."[358] The titles to Stock's books speak for themselves concerning what germ-line engineering would do to the human race. The name of one is *Redesigning Humans: Our Inevitable Genetic Future* and another is *Metaman: The Merging of Humans and Machines into a Global Superorganism.*

Besides the short list above, additional areas of concern where readers may wish to become well advised on the pros and cons of enhancement technology include immortalism, postgenderism, augmented reality, cryonics, designer babies, neurohacking, mind uploading, neural implants, xenotransplantation, reprogenetics, rejuvenation, radical life extension, and more.

Heaven and Hell Scenarios

While positive advances either already have been or will come from some of the science and technology fields we have discussed, learned men like Prof. Francis Fukuyama, in his book, *Our Posthuman Future: Consequences of the Biotechnology Revolution,* warn that unintended consequences resulting from what mankind has now set in motion represents the most dangerous time in earth's history, a period when exotic technology in the hands of transhumanist ambitions could forever alter what it means to be human. To those who would engineer a transhuman future, Fukuyama warns of a dehumanized "hell scenario" in which we "no longer struggle, aspire, love, feel pain, make difficult moral choices, have families, or do any of the things that we traditionally associate with being human." In this ultimate identity crisis, we would "no longer have the characteristics that give us human dignity" because, for one thing, "people dehumanized à la *Brave New World*…don't know that they are dehumanized, and, what is worse, would not care if they knew. They are, indeed, happy slaves with a slavish happiness."[359] The "hell scenario" envisioned by Fukuyama is but a beginning to what other intelligent thinkers believe could go wrong.

On the other end of the spectrum and diametrically opposed to Fukuyama's conclusions is an equally energetic crowd that subscribes to a form of technological utopianism called the "heaven scenario." Among this group, a "who's who" of transhumansist evangelists such as Ray Kurzweil, James Hughes, Nick Bostrom, and Gregory Stock see the dawn of a new Age of Enlightenment arriving as a result of the accelerating pace of GRIN technologies. As with the eighteenth-century Enlightenment in which intellectual and scientific reason elevated the authority of scientists over priests, techno-utopians believe they will triumph over prophets of doom by "stealing fire from the gods, breathing life into inert matter, and gaining immortality. Our efforts to become something more than human have a long and distinguished genealogy. Tracing the history of those efforts illuminates human nature. In every civilization, in every era, we have given the gods no peace."[360] Such men are joined in their quest for godlike constitutions by a growing list of official US departments that dole out hundreds of millions of dollars each year for science and technology research. The National Science Foundation and the United States Department of Commerce anticipated this development over a decade ago, publishing the government report *Converging Technologies for Improving Human Performance*—complete with diagrams and bullet points—to lay out the blueprint for the radical evolution of man and machine. Their vision imagined that, starting around the year 2012, the "heaven scenario" would begin to be manifested and quickly result in (among other things):

- The transhuman body being "more durable, healthy, energetic, easier to repair, and resistant to many kinds of stress, biological threats, and aging processes."
- Brain-machine interfacing that will "transform work in factories, control automobiles, ensure military superiority, and enable new sports, art forms and modes of interaction between people."
- "Engineers, artists, architects, and designers will experience tremendously expanded creative abilities," in part through "improved understanding of the wellspring of human creativity."
- "Average persons, as well as policymakers, will have a vastly improved awareness of the cognitive, social, and biological forces operating their lives, enabling far better adjustment, creativity, and daily decision making.... Factories of tomorrow will be organized" around "increased human-machine capabilities."[361]

Beyond how human augmentation and biological reinvention would spread into the wider culture following 2012 (the same date former counter-terrorism czar, Richard Clark, in his book, *Breakpoint,* predicted serious GRIN rollout), the government report detailed the *especially* important global and economic aspects of genetically superior humans acting in superior ways, offering how, as a result of GRIN leading to techno-sapien DNA upgrading, brain-to-brain interaction, human-machine interfaces, personal sensory device interfaces, and biological war fighting systems, "The twenty-first century could end in world peace, universal prosperity, and evolution to a higher level [as] humanity become[s] like a single, transcendent nervous system, an interconnected 'brain' based in new core pathways of society." The first version of the government's report asserted that the only real roadblock to this "heaven scenario" would be the "catastrophe" that would be unleashed if society fails to employ the technological opportunities available to us now. "We may not have the luxury of delay, because the remarkable economic, political and even violent turmoil of recent years implies that the world system is unstable. If we fail to chart the direction of change boldly, we may become the victims of unpredictable catastrophe."[362] This argument parallels what is currently echoed in military corridors, where sentiments hold that failure to commit resources to develop GRIN as the next step in human and technological evolution will only lead to others doing so ahead of us and using it for global domination.

Not everybody likes the "heaven scenario" imperative, and from the dreamy fantasies of *Star Trek* to the dismal vision of Aldous Huxley's *Brave New World*, some have come to believe there are demons hiding inside transhumanism's mystical (or mythical?) "Shangri-la."

"Many of the writers [of the government report cited above] share a faith in technology which borders on religiosity, boasting of miracles once thought to be the province of the Almighty," write the editors of *The New Atlantis: A Journal of Technology and Society*. "[But] without any serious reflection about the hazards of technically manipulating our brains and our consciousness…a different sort of catastrophe is nearer at hand. Without honestly and seriously assessing the consequences associated with these powerful new [GRIN] technologies, we are certain, in our enthusiasm and fantasy and pride, to rush headlong into disaster."[363]

Few people would be more qualified than computer scientist Bill Joy to annunciate these dangers, or to outline the "hell scenario" that could unfold as a result of GRIN. Yet it must have come as a real surprise to some of those who remembered him as the level-headed Silicon Valley scientist and co-founder of Sun Microsystems (SM) when, as chief scientist for the corporation, he released a vast and now-famous essay, "Why the Future Doesn't Need Us," arguing how GRIN would threaten in the very near future to obliterate mankind. What was extraordinary about Joy's prophecy was how he saw himself—and people like him—as responsible for building the very machines that "will enable the construction of the technology that may replace our species."

"From the very moment I became involved in the creation of new technologies, their ethical dimensions have concerned me," he begins. But it was not until the autumn of 1998 that he became "anxiously aware of how great are the dangers facing us in the twenty-first century." Joy dates his "awakening" to a chance meeting with Ray Kurzweil, whom he talked with in a hotel bar during a conference at which they both spoke. Kurzweil was finishing his manuscript for *The Age of Spiritual Machines* and the powerful descriptions of sentient robots and near-term enhanced humans left Joy taken aback, "especially given Ray's proven ability to imagine and create the future," Joy wrote. "I already knew that new technologies like genetic engineering and nanotechnology were giving us the power to remake the world, but a realistic and imminent scenario for intelligent robots surprised me."

Over the weeks and months following the hotel conversation, Joy puzzled over Kurzweil's vision of the future until finally it dawned on him that genetic engineering, robotics, artificial intelligence, and nanotechnology posed "a different threat than the technologies that have come before. Specifically, robots, engineered organisms, and nanobots share a dangerous amplifying factor: They can self-replicate. A bomb is blown up only once—but one bot can become many, and quickly get out of control." The unprecedented threat of self-replication particularly burdened Joy because, as a computer scientist, he thoroughly understood the concept of out-of-control replication or viruses leading to machine systems or computer networks being disabled. Uncontrolled self-replication of nanobots or engineered organisms would run "a much greater risk of substantial damage in the physical world," Joy concluded before adding his deeper fear:

> What was different in the twentieth century? Certainly, the technologies underlying the weapons of mass destruction (WMD)—nuclear, biological, and chemical (NBC)—were powerful, and the weapons an enormous threat. But building nuclear weapons required...highly protected information; biological and chemical weapons programs also tended to require large-scale activities.
>
> The twenty-first-century technologies—genetics, nanotechnology, and robotics...are so powerful that they can spawn

whole new classes of accidents and abuses. Most dangerously, for the first time, these accidents and abuses are widely within the reach of individuals or small groups. They will not require large facilities or rare raw materials. Knowledge alone will enable the use of them.

Thus we have the possibility not just of weapons of mass destruction but of knowledge-enabled mass destruction (KMD), this destructiveness hugely amplified by the power of self-replication.

I think it is no exaggeration to say we are on the cusp of the further perfection of extreme evil, an evil whose possibility spreads well beyond that which weapons of mass destruction bequeathed to the nation states, on to a surprising and terrible empowerment.[364]

Joy's prophecy about self-replicating "extreme evil" as an imminent and enormous transformative power that threatens to rewrite the laws of nature and permanently alter the course of life as we know it was frighteningly revived this year in the creation of Venter's "self-replicating" Synthia species (Venter's description). Parasites such as the mycoplasma mycoides that Venter modified to create Synthia can be resistant to antibiotics and acquire and smuggle DNA from one species to another, causing a variety of diseases. The dangers represented by Synthia's self-replicating parasitism has thus refueled Joy's opus and given experts in the field of counter-terrorism sleepless nights over how extremists could use open-source information to create a Frankenstein version of Synthia in fulfillment of Carl Sagan's *Pale Blue Dot*, which Joy quoted as, "the first moment in the history of our planet when any species, by its own voluntary actions, has become a danger to itself." As a dire example of the possibilities this represents, a genetically modified version of mouse pox was created not long ago that immediately reached 100 percent lethality. If such pathogens were unleashed into population centers, the results would be catastrophic. This is why Joy and others were hoping a few years ago that a universal moratorium or voluntary relinquishment of GRIN developments would be initiated by national laboratories and governments. But the genie is so far out of the bottle today that even college students are attending annual synthetic biology contests (such as the International Genetically Engineered Machine Competition, or iGEM) where nature-altering witches' brews are being concocted by the scores, splicing and dicing DNA into task-fulfilling living entities. For instance, the iGEM 2009 winners built "E. chromi"—a programmable version of the bacteria that often leads to food poisoning, *Escherichia coli* (commonly abbreviated *E. coli*). A growing list of similar DNA sequences are readily available over the Internet, exasperating security experts who see the absence of universal rules for controlling what is increasingly available through information networks as threatening to unleash a "runaway sorcerer's apprentice" with unavoidable biological fallout. Venter and his collaborators say they recognize this danger—that self-replicating biological systems like the ones they are building—hold peril as well as hope, and they have joined in calling on Congress to enact laws to attempt to control the flow of information and synthetic "recipes" that could provide lethal new pathogens for terrorists. The problem, as always, is getting all of the governments in the world to voluntarily follow a firm set of ethics or rules. This is wishful thinking at best. It is far more likely the world is racing toward what Joel Garreau was first to call the "hell scenario"—a moment in which human-driven GRIN technologies place earth and all its inhabitants on course to self-eradication.

Ironically, some advocates of posthumanity are now using the same threat scenario to advocate *for* transhumanism as the best way to deal with the inevitable extinction of mankind via GRIN. At the global interdisciplinary institute Metanexus (www.metanexus.net/), Mark Walker, assistant professor at New Mexico State University (who holds the Richard L. Hedden of Advanced Philosophical Studies Chair) concludes like Bill Joy that "technological advances mean that there is a high probability that a human-only future will end in extinction." From this he makes a paradoxical argument:

In a nutshell, the argument is that even though creating posthumans may be a very dangerous social experiment, it is even more dangerous not to attempt it....

I suspect that those who think the transhumanist future is risky often have something like the following reasoning in mind: (1) If we alter human nature then we will be conducting an experiment whose outcome we cannot be sure of. (2) We should not conduct experiments of great magnitude if we do not know the outcome. (3) We do not know the outcome of the transhumanist experiment. (4) So, we ought not to alter human nature.

The problem with the argument is.... Because genetic engineering is already with us, and it has the potential to destroy civilization and create posthumans, we are already entering uncharted waters, so we must experiment. The question is not whether to experiment, but only the residual question of which social experiment will we conduct. Will we try relinquishment? This would be an unparalleled social experiment to eradicate knowledge and technology. Will it be the steady-as-she-goes experiment where for the first time governments, organizations and private citizens will have access to knowledge and technology that (accidently or intentionally) could be turned to civilization ending purposes? Or finally, will it be the transhumanist social experiment where we attempt to make beings brighter and more virtuous to deal with these powerful technologies?

I have tried to make at least a *prima facie* case that transhumanism promises the safest passage through twenty-first century technologies.[365]

The authors of this book believe the "brighter and more virtuous beings" Professor Walker and others are arguing for possess supernatural elements and that the *spirit* behind the transhumanist nightmare will put the "hell" in the "hell scenario" sooner than most comprehend.

The Transhuman New Face of Spiritual Warfare

13

By Thomas and Nita Horn

Publisher's Note This and the preceding chapter are included in *The Milieu* because these opinions and research by Dr. Thomas and Nita Horn—first published in their book *Forbidden Gates* (Defender Publishing, 2011)—is specifically referenced by S. Jonathon O'Donnell in his peer-reviewed treatise on "Horn's Milieu" titled: *Secularizing Demons: Fundamentalist Navigations in Religion and Secularity.*

> If the U.S. [today] has a national religion, the closest thing to it is faith in technology.
> —Scott Keeter, director of survey research for the
>   Pew Research Center

> Yet again humankind seems ready to plunge headlong into another human, or demonic, contrivance promising salvation and eternal happiness for all. This time the Faustian bargain is being struck with technology, what John McDermott referred to as the "opiate of the intellectuals."
> —C. Christopher Hook, MD

When the stars align, Cthulhu will rise again to resume His dominion over the Earth, ushering in an age of frenzied abandon. Humankind will be "free and wild and beyond good and evil, with laws and morals thrown aside and all men shouting and killing and reveling in joy."

—Transhumanist Mark Dery, celebrating the rise of
    H. P. Lovecraft's cosmic monster

On July 20, 2010, the *New York Times* ran a feature article introducing a new nonprofit organization called the Lifeboat Foundation.[366] The concept behind the group is simple yet disturbing. Protecting people from threats posed by potentially catastrophic technology—ranging from artificial intelligence running amok to self-replicating nanobots—represents an emerging opportunity for designing high-tech "shields," and lots of them, to protect mankind this century.

"For example," the article says, "there's talk of a Neuroethics Shield to prevent abuse in the areas of neuropharmaceuticals, neurodevices, and neurodiagnostics. Worse cases include enslaving the world's population or causing everyone to commit suicide.

"And then there's a Personality Preserver that would help people keep their personalities intact and a Nano Shield to protect against overly aggressive nanocreatures."

If the Lifeboat Foundation sounds like a storehouse for overreacting geeks or even outright nut jobs, consider that their donors involve Google, Hewlett-Packard, Sun Microsystems, and an impressive list of industry and technology executives, including names on their advisory boards like Nobel laureate and Princeton University Prof. Eric Maskin.

What the development of such enterprising research groups illustrates is that even if one does not believe speculation from the previous chapters suggesting mind-bending concepts like Nephilim being resurrected into posthuman bodies via GRIN technology, all of society—regardless of religious or secular worldviews—should consider that what we are doing now through genetic modification of living organisms and the wholesale creation of new synthetic life-forms is either a violation of the divine order (biblical creation, such as the authors of this book believe) or chaos upon natural evolution, or both. The road we have started down is thus wrought with unknown perils, and the Lifeboat Foundation is correct to discern how the transhuman era may abruptly result in the need for "shields" to protect earth species from designer viruses, nanobugs, prion contamination, and a host of other clear and present dangers. Part of the obvious reasons behind this is, in addition to the known shortcomings of biotechnology corporations and research facilities to remain impartial in their safety reviews (they have a vested interest in protecting approval and distribution of their products), futurist think tanks such as the Lifeboat Foundation understand that the phrase, "those who fail to learn from history are doomed to repeat it" is axiomatic for a reason. Human nature has a clear track record of developing defense mechanisms only after natural or manufactured threats have led to catastrophe. We humans seem doomed to learn from our mistakes far more often than from prevention. Consider how nuclear reactors were forced to become safer only after the Chernobyl disaster, or how a tsunami warning system was developed by the United Nations following 230,000 people being killed by a titanic wave in the Indian Ocean. This fact of human nature portends an especially ill wind for mankind when viewed against the existential threats of biological creations, artificial intelligence systems, or geoengineering of nature, which carry the potential not only of backfiring but of permanently altering the course of humanity. "Our attitude throughout human history has been to experience events like these and then to put safeguards in place," writes Prof. Nick Bostrom. "That strategy is completely futile with existential risks [as represented in GRIN tech]. By definition, you don't get to learn from experience. You only have one chance to get it right."[367] Because of the truly catastrophic threat thus posed by mostly unregulated GRIN advances this century, Richard Posner, a US appeals court judge and author of the book *Catastrophe: Risk and Response*, wants "an Office of

Risk and Catastrophe set up in the White House. The office would be charged with identifying potentially dangerous technologies and calling in experts to inform its own risk assessment." The problem right now, Posner adds, "is that no single government department takes responsibility for these kinds of situations."[368] Not surprisingly, many transhumanists contest Posner's idea, saying it represents just another unnecessary bureaucracy that would stand in the way of scientific progress.

Yet of greater significance and repeatedly missing from such secular considerations is what the authors of this book believe to be the more important element: supernaturalism and spirituality. Beyond the material ramifications of those threats posed by the genetics revolution is something most scientists, engineers, and bioethicists fail to comprehend—that man is not just a series of biological functions. We are spirit and soul and vulnerable to spiritual, not just environmental, dangers. Thus the "shields" that the Lifeboat Foundation is working on will only protect us so far. We will need *spiritual shields* too as GRIN raises those bigger issues of how human-transforming enhancements may alter our very souls (says Joel Garreau) as well as hundreds of immediate new challenges that Christians, families, and ministries will be facing.

It is an understatement to say that technology often works hand in hand with unseen forces to challenge our faith or open new channels for spiritual warfare. This has been illustrated in thousands of ways down through time—from the creation of Ouija boards for contacting the spirit world to online pornography gateways. But the current course upon which GRIN technology and transhumanist philosophy is taking mankind threatens to elevate the reality of these dangers to quantitatively higher levels. Some of the spiritual hazards already surfacing as a result of modern technology include unfamiliar terms like "i-Dosing," in which teens get "digitally high" by playing specific Internet videos through headphones that use repetitive tones to create binaural beats, which have been shown in clinical studies to induce particular brain-wave states that make the sounds appear to come from the center of the head. Shamans have used variations of such repetitive tones and drumming to stimulate and focus the "center mind" for centuries to make contact with the spirit world and to achieve altered states of consciousness.

More broadly, the Internet itself, together with increasing forms of electronic information-driven technology, is creating a new kind of addiction by "rewiring our brains," says Nora Volkow, world-renowned brain scientist and director of the National Institute of Drug Abuse. The lure of "digital stimulation" can actually produce dopamine releases in the brain that affect the heart rate and blood pressure and lead to drug-like highs and lows. As bad, the addictive craving for digital stimulation is leading to the electronic equivalent of Attention Deficit Disorder (ADD) among a growing population in which constant bursts of information and digital stimulation undermine one's ability to focus—especially in children, whose brains are still developing and who naturally struggle to resist impulses or to neglect priorities. A growing body of literature is verifying this e-connection to personality fragmentation, cyber relationships over personal ones, and other psychosocial issues. Volkow and other researchers see these antisocial trends leading to widespread diminished empathy between people—which is essential to the human condition—as a result of humans paying more and more attention to iPads, cell phones, and computer screens than to each other, even when sitting in the same room. New research shows this situation becoming an electronic pandemic as people escalate their detachment from traditional family relationships while consuming three times as much digital information today as they did in 2008, checking e-mails, texting thirty-seven times per hour, and spending twelve hours per day on average taking in other e-media.

How brain-machine interfacing will multiply this divide between human-to-human relationships versus human-machine integration should be of substantial concern to readers for several reasons, including how 1) the Borgification of man will naturally exasperate the decline of the family unit and interpersonal relationships upon which society has historically depended; 2) the increase of euphoric cybernetic addiction will multiply as cerebral stimulation of the brain's pleasure centers is added to existing natural senses—sight, hearing, taste, smell, and touch; and 3) the threat of computer viruses or hijackers disrupting enhanced human neural or cognitive pathways will develop as cyber-enhanced individuals evolve. To illustrate the latter, Dr. Mark Gasson, from the School of Systems Engineering at the University of Reading in the United Kingdom, intentionally contaminated an implanted microchip in his hand that allows him biometric entry through security doors and that also communicates with his cell phone and other external devices. In the experiment, Dr. Gasson (who agrees that the next step in human evolution is the transhuman vision of altered human biology integrated with machines) was able to show how the computer virus he infected himself with spread to external computer systems in communication with his microchip. He told BBC News, "With the benefits of this type of technology come risks. We [will] improve ourselves…but much like the improvements with other technologies, mobile phones for example, they become vulnerable to risks, such as security problems and computer viruses."[369]

Such threats—computer viruses passing from enhanced humans to enhanced humans via future cybernetic systems—are the tip of the iceberg. The real danger, though it may be entirely unavoidable for some, will be the loss of individuality, anonymity, privacy, and even free will as a result of cybernetic integration. Dr. Christopher Hook contends, "If implanted devices allow the exchange of information between the biological substrate and the cybernetic device," such a device in the hippocampus (the part of the brain involved in forming, storing, and processing memory) for augmenting memory, for instance, "would be intimately associated with the creation and recall of memories as well as with all the emotions inherent in that process. If this device were… to allow the importation of information from the Internet, could the device also allow the memories and thoughts of the individual to be downloaded or read by others? In essence, what is to prevent the brain itself from being hacked [or externally monitored]? The last bastion of human privacy, the brain, will have been breached."[370]

Despite these significant ethical and social dangers, industry and government interest in the technological dream of posthumanism, as documented earlier in this book, is more than *laissez-faire*. The steady migration toward the fulfillment of biologically and cybernetically modified humans combined with corporate and national investments will predictably fuse this century, ultimately leading to strong cultural forces compelling all individuals to get "plugged in" to the grid. Whoever resists will be left behind as inferior Luddites (those who oppose new technology), or worse, considered enemies of the collectives' progress, as in de Garis' nightmarish vision in the *Artilect War* or former counter-terrorism czar Richard Clark's *Breakpoint,* which depicts those who refuse technological enhancement as "terrorists."

According to the work *Human Dignity in the Biotech Century,* this pressure to become enhanced will be dramatic upon people in all social strata, including those in the middle class, law, engineering, finance, professional fields, and the military, regardless of personal or religious views:

> Consider…whether the military, after investing billions in the development of technologies to create the cyborg soldier… would allow individual soldiers to decline the enhancements because of religious or personal qualms. It is not likely. Individuals may indeed dissent and decline technological augmentation, but such dissenters will find job options increasingly scarce.
>
> Because the network of cyborgs will require increasing levels of cooperation and harmonious coordination to further improve efficiency, the prostheses will continue to introduce means of controlling or modulating emotion to promote these values. Meanwhile, the network is increasingly controlled by central planning structures to facilitate harmony and efficiency. While everyone still considers themselves fully autonomous, in reality behavior is more and more tightly controlled. Each step moves those who are cybernetically augmented toward becoming like the Borg, the race of cybernetic organisms that inhabit the twenty-sixth century of the *Star Trek* mythology. The Borg,

once fully human, become "assimilated" by the greater collective mind, losing individuality for the good of the whole.[371]

Lest anyone think the writers of *Human Dignity in the Biotech Century* are overly paranoid, consider that NBIC (Nanotechnology, Biotechnology, Information Technology, and Cognitive Science) director Mihail Roco, in the US government report, *Converging Technologies for Improving Human Performance*, wrote:

> Humanity would become like a single, distributed and interconnected "brain" based in new core pathways in society.... A networked society of billions of human beings could be as complex compared to an individual being as a human being is to a single nerve cell. From local groups of linked enhanced individuals to a global collective intelligence, key new capacities would arise from relationships arising from NBIC technologies.... Far from unnatural, such a collective social system may be compared to a larger form of biological organism.... We envision the bond of humanity driven by an *interconnected virtual brain* of the Earth's communities searching for intellectual comprehension and conquest of nature.[372]

Nowhere will the struggle to resist this human biological alteration and machine integration be more immediate than in those religious homes where transhumanism is seen as an assault on God's creative genius, and where, as a result, people of faith seek to maintain their humanity. Yet the war against such believers is poised to emerge over the next decade as much from inside these homes and families as it will from external social influences.

As a simple example, flash forward to the near future when much of the technology previously discussed—factually based on emerging technologies and anticipated time frames—is common. Your tenth-grade daughter, Michelle, walks in from a first day at a new school.

"Well, how did it go, Honey?" you ask with a smile.

"It was okay," she says, "though the kids here are even smarter than at the last school." But then she pauses. She knows begging to be enhanced like most of her classmates will only lead to more arguing—common between you two on this subject. How can she make you understand what it's like even trying to compete with the transhumans? The fact that most of the student body, students who are half her age, will graduate from college *summa cum laude* with IQs higher than Einstein's by the time she even enters is a ridiculous and unnecessary impediment, she feels. She can't understand it. You've seen the news, the advertising, the *H+* magazines articles and television specials outlining the advantages of enhancement. Even the family doctor tried to convince you. But it will probably take a visit from Child Welfare Services, which in the US is soon to follow the European model where, starting in 2019, parents whose children went without basic modifications were charged with neglect and had their kids put in foster homes. She just wishes it wouldn't come to that. If only you could be like those Emergent Christians 2.0 whose techno-theology arose during the early enhancement craze of 2016–2018, based on a universalist imperative for "perfectionist morality" and the Christian duty to be "healers and perfecters" as opposed to the "bio-Luddite theology" of your outdated religious "divine order" concept, which only serves to keep people like her at disadvantage. That's why she gave you the school report compiled by Prof. Joel Garreau describing the average high school pupil today, so you could understand how her classmates:

- Have amazing thinking abilities. They're not only faster and more creative than anybody she's ever met, but faster and more creative than anybody she's ever imagined.
- They have photographic memories and total recall. They can devour books in minutes.
- They're beautiful, physically. Although they don't put much of a premium on exercise, their bodies are remarkably ripped.

- They talk casually about living a long time, perhaps being immortal. They're always discussing their "next lives." One fellow mentions how, after he makes his pile as a lawyer, he plans to be a glassblower, after which he wants to become a nanosurgeon.
- One of her new friends fell while jogging, opening up a nasty gash on her knee. Your daughter freaked, ready to rush her to the hospital. But her friend just stared at the gaping wound, focusing her mind on it. Within minutes, it simply stopped bleeding.
- This same friend has been vaccinated against pain. She never feels acute pain for long.
- These new friends are always connected to each other, sharing their thoughts no matter how far apart, with no apparent gear. They call it "silent messaging." It seems like telepathy.
- They have this odd habit of cocking their head in a certain way whenever they want to access information they don't yet have in their own skulls—as if waiting for a delivery to arrive wirelessly…which it does.
- For a week or more at a time, they don't sleep. They joke about getting rid of their beds, since they use them so rarely.[373]

Even though these enhanced students treat her with compassion and know that she is biologically and mentally handicapped by no fault of her own, she hates it when they call her a "Natural." It feels so condescending. And then, at the last school, there was that boy she wanted to date, only to discover it was against the informed-consent regulations passed by the Department of Education two years ago restricting romantic relationships between "Naturals" and the "Enhanced." She could have crawled into a hole, she was so embarrassed. But she's decided not to fight you anymore about it. Next year she will be eighteen years old and has been saving her money. With the federal Unenhanced Student Aid programs administered by the US Department of Education and the United Naturals Student Fund (UNSF) that provides financial assistance and support for "Disaugmented American Students," grades pre-kindergarten to twelve, whose motto is "An augmented mind is a terrible thing to waste," she'll have enough for Level 1 Genetic Improvement plus a couple of toys like Bluetooth's new extracranial cybernetic communicator. It's not much, but it's a start, and though you will tell her that her brain-machine interface, and especially her genetic upgrade, makes her—as well as any kids she has in the future—inhuman, according to the school's genetic guidance counselor, there will be nothing you can do to legally stop her.

The Devil Is in the Details

As transhumanist philosophy and GRIN technology become integrated within society and national and private laboratories with their corporate allies provide increasingly sophisticated arguments for its widest adoption, those of us who treasure the meaning of life and human nature as defined by Judeo-Christian values will progressively find ourselves engaged in deepening spiritual conflicts over maintaining our humanity in the midst of what the authors believe is fundamentally a supernatural conflict.

Just as the fictional exercise with the seventeen-year-old "Michelle" above illustrates, intensifying techno-spiritual issues, which Christian families will face this century, will escalate simultaneously at both spiritual and scientific levels. This material/immaterial struggle, which philosopher and theologian Francis Schaeffer once described as always at war "in the thought-world," is difficult for some to grasp. The idea that human-transforming technology that mingles the DNA of natural and synthetic beings and merges man with machines could somehow be used or even inspired by *evil supernaturalism* to foment destruction within the material world is for some people so exotic as to be inconceivable. Yet nothing should be more fundamentally clear, as students of spiritual warfare will understand. We are body (physical form), mind (soul, will, emotions), and spirit, thus everything in the material and immaterial world has potential to influence our psychosomatic existence and decisions. "There is no conflict in our lives that is strictly a spiritual issue," writes Robert Jeffress in his book, *The Divine Defense*. This is because "there is never a time when the spirit is divorced from the body. Likewise, there is no turmoil in our lives that is solely psychological or physical, because our spirit, along with God's Spirit within us and demonic spirits around us, is always present as well."[374] Jeffress' point that material stimulus cannot be divorced from spiritual conditions conveys why the Bible is so concerned with the antitheses of transhumanism; the integrity of our bodies and minds. The goal is to bring both into obedience to Christ (2 Chronicles 10:5) because this is where the battle is first fought and won. No marriage breakup ever transpired that did not start there—no murder, no theft, no idolatry—but that the contest was staged in the imagination, then married to the senses, and the decision to act given to the victor.

How technology is now poised to raise this mind-body-spirit game is hidden in the shadows of the National Institute of Health and DARPA, which for more than three decades have invested hundreds of millions of dollars not only designing new DNA constructs but crafting arrays of microelectrodes, supercomputers, and algorithms to analyze and decipher the brain's neural code, the complex "syntax" and communication rules that transform electrical neuron pulses in the brain into specific digital and analog information that we ultimately perceive as decisions, memories, and emotions. Understanding how this secret brain language functions, then parsing it down into digital computer code (strings of ones and zeros) where it can be reassembled into words and commands and then manipulated is at the center of military neurobiology, artificial intelligence research, and cybernetics.

While significant studies in neurosciences have been conducted with "neuro-prostheses" in mind to help the handicapped—for instance, the artificial cochlear implant that approximately 188,000 people worldwide have received thus far—DARPA "is less interested in treating the disabled than in enhancing the cognitive capacities of soldiers," says former senior writer at *Scientific American*, John Horgan. "DARPA officials have breached the prospect of cyborg warriors downloading complex fighting procedures directly into their brains, like the heroes of the Matrix," and has "interest in the development of techniques that can survey and possibly manipulate the mental processes of potential enemies [by] recording signals from the brains of enemy personnel at a distance, in order to 'read their minds and to control them.'"[375] Because what develops within military technology eventually migrates into the broader culture, where it is quickly embraced for competitive or mutual advantages, the ramifications of neurobiology have not escaped international interests in both public and private agencies. Entire fields of research are now under development worldwide based on the notion that breakthroughs will provide unprecedented opportunities for reading, influencing, and even controlling human minds this century. The implications from this field are so staggering that France, in 2010, became the first nation to establish a behavioral research unit specifically designed to study and set "neuropolicy" to govern how such things as "neuromarketing" (a new field of marketing that analyzes consumers' sensorimotor and cognitive responses to stimuli in order to decode what part of the brain is telling consumers to make certain buying decisions) may be used in the future to access unconscious decision-making elements of the brain to produce desired responses. This precedent for government neuropolicy comes not a second too soon, as the world's largest semiconductor chip maker, Intel Corp., wants brain communicators on the market and "in its customers' heads" before the year 2020. In what can only be described as *Matrix* creep, researchers at Intel Labs Pittsburgh are designing what it bets will be "the next big thing"—brain chips that allow consumers to control a host of new electronic and communication gadgets by way of neural commands. Developers at Toyota and the University of Utah are also working on brain transmitters, which they hope will contribute to building a global "hive mind."

From these developments comes the distant groaning of a "fearful unknown" in which the architecture of the human brain—as transformed by current and future cybernetic inventions—begins to act in ways that borderline the supernatural. Consider experimental telepathy, which involves mind-to-mind thought transference that allows people to communicate without the use of speaking audibly. Most do not know that Hans Berger, the inventor of electroencephalography (EEG, the recording of electrical activity along the scalp produced by the firing of neurons within the brain) was a strong believer in psychic phenomena and wanted to decode brain signals in order to establish nonverbal transmission between people. GRIN technology proposes to fulfill his dream.

Another example is telekinesis (psychokinesis), which involves the movement or manipulation of physical matter via direct influence of the mind. As incredible as it may seem, both this idea and the one above are under research by DARPA and other national laboratories as no pipe dream. Such brain-to-brain transmission between distant persons as well as mind-to-computer communication was demonstrated last year at the University of Southampton's Institute of Sound and Vibration Research using electrodes and an Internet connection. The experiment at the institute went farther than most brain-to-machine interfacing (BMI) technology thus far, actually demonstrating brain-to-brain (B2B) communication between persons at a distance. Dr. Christopher James, who oversaw the experiment, commented: "Whilst BCI [brain-computer interface] is no longer a new thing and person-to-person communication via the nervous system was shown previously in work by Prof. Kevin Warwick from the University of Reading, here we show, for the first time, true brain to brain interfacing. We have yet to grasp the full implications of this." The experiment allowed one person using BCI to transmit thoughts, translated as a series of binary digits, over the Internet to another person whose computer received the digits and transmitted them to the second user's brain.[376]

The real danger is how these accomplishments within human-mind-to-synthetic intelligence may take the proverbial "ghost in the machine" where no *modern man* has gone before, bridging a gap between unknown entities (both virtual and real), perhaps even inviting takeover of our species by malevolent intelligence. Note that the experiments above are being conducted at Southampton's Institute of Sound and Vibration Research. Some years ago, scientist Vic Tandy's research into sound, vibration frequencies, and eyeball resonation led to a thesis (actually titled "Ghosts in the Machine") that was published in the *Journal of the Society for Psychical Research*. Tandy's findings outlined what he thought were "natural causes" for particular cases of specter materialization. Tandy found that 19-Hz standing air waves could, under some circumstances, create sensory phenomena in an open environment suggestive of a ghost. He actually produced a frightening manifested entity resembling contemporary descriptions of "alien grays." A similar phenomenon was discovered in 2006 by neurologist Olaf Blanke of the Brain Mind Institute in Lausanne, Switzerland, while working with a team to discover the source of epileptic seizures in a young woman. They were applying electrical currents through surgically implanted electrodes to various regions of her brain, when upon reaching her left temporo-parietal junction (TPJ, located roughly above the left ear) she suddenly reported feeling the presence of a shadow person standing behind her. The phantom started imitating her body posture, lying down beneath her when she was on the bed, sitting behind her, and later even attempting to take a test card away from her during a language exercise. While the scientists interpreted the activity as a natural, though mysterious, biological function of the brain, is it possible they were actually discovering gateways of perception into the spirit world that were closed by God following the Fall of Man? Were Tandy's "ghost" and Blanke's "shadow person" *living unknowns?* If so, is it not troubling that advocates of human-mind-to-machine intelligence may produce permanent conditions similar to Tandy and Blanke's findings, giving rise to simulated or real relationships between humans and "entities"? At the thirteenth European Meeting on Cybernetics and Systems Research at the University of Vienna in Austria, an original paper submitted by Charles Ostman seemed to echo this possibility:

As this threshold of development is crossed, as an index of our human/Internet symbiosis becoming more pronounced, and irreversible, we begin to develop communication modalities which are quite "nonhuman" by nature, but are "socio-operative" norms of the near future. Our collective development and deployment of complex metasystems of artificial entities and synthetic life-forms, and acceptance of them as an integral component of the operational "culture norm" of the near future, is in fact the precursory developmental increment, as an enabling procedure, to gain effective communicative access to *a contiguous collection of myriad "species" and entity types (synthetic and "real") functioning as process brokeraging agents.*[377]

A similar issue that "pinged" in our memories from past experience with exorcism and the connection between *sound resonance* and contact with supernaturalism has to do with people who claim to have become possessed or "demonized" after attempting to open mind gateways through vibratory chanting at New Age vortices or "Mother Earth" energy sites such as Sedona, Arizona. When we queried www.RaidersNewsNetwork. com resident expert Sue Bradley on this subject, asking if she believed a connection could exist between acoustics, harmonics, sound resonance, and spirit gateways, she e-mailed this lengthy and shocking response:

Tom and Nita:
    From the ancients to the New Age, resonance and harmonics have long been recognized as vehicles of communication and manifestation. Ancient rock outcroppings, sacred temples, and monuments have for millennia been used as gathering places for the so-called spiritually enlightened. Through recent understanding of quantum entanglement and the high energy physics of sound and light, both with adaptable vibratory characteristics, these popular sites for gatherings with ritual chants and offerings, often employing ancient spells and mathemati-

cal harmonic codes in various sets of tandem frequencies, may well have measurable and far greater esoteric effects than even recently believed.

Note what New Ager and modern shaman Zacciah Blackburn of *Sacred Sound, Sound the Codes* says he came in contact with at such sites:

> It is not mere coincidence many of the ancient stone temples of the world were made with crystalline embedded stones, such as granite, which are known for their properties to pass or store energy.... Through Sacred Sound and awareness practices, the unseen "wisdom keepers" and guardians of these sacred temples have communed with me, and showed me how to hold frequency of awareness in the heart and mind, and combine them into sound codes to create a "key" which opens the "libraries" of these temples of ancient star beings and wisdom keepers to the modern way traveler whom comes with pure intent.[378]

With this in mind, also consider how the word "ear" appears 120 times in Scripture, "ears" 151 times, and is important with regard to *sound connected to spiritual hearing.* First used in Exodus 15:26, the ear is linked to a covenant relationship for those that *hearken* to the voice of the Lord and keep His statutes. The *right ear* is repeatedly described in the Levitical instructions: "Then shalt thou kill the ram, and take of his blood, and put it upon the tip of the right ear of Aaron, and upon the tip of the right ear of his sons" (Exodus 29:20; [Leviticus 8:23 and 24; 14:14, 17, 25, and 28]).

Subsequent references to the ear and hearing are presented as petitions *to* God from His servants as well as *from* God as counsel, forewarning and rebukes.

The ear as a spiritual gateway termed "Ear-Gate" first appeared in English usage through an allegory penned by John Bunyan in 1682. Bunyan's classic, *The Pilgrim's Progress,* was the most widely read and translated book in the English language apart from the Bible: it was also an educational staple and considered to be required reading in the U.S. from colonial times through World War II. While *The Pilgrim's Progress* allegorizes the encounters and obstacles of a man seeking salvation, Bunyan's *The Holy War* or *The Losing and Taking Again of the Town of Mansoul* recounts the cosmic conflict for the souls of mankind with Peretti-like descriptions and precision.

The town of Mansoul, designed in the image of the almighty, *Shaddai*, is the target of the deceptive and malevolent giant, *Diabolus*. Mansoul is a city of five gates: the Ear-Gate, Eye-Gate, Mouth-Gate, Nose-Gate and Feel-Gate. The first and most strategic gate is the first gate breached: the Ear-Gate.

Nineteenth-century theologian, Rev. Robert Maguire, comments on the importance of the Ear-Gate:

> This was the gate of audience, and through this gate the words of the tempter must penetrate, if the temptation is to be successful. Into the ears of our first mother did the wily serpent whisper the glozing words of his seductive wiles and through the Ear-Gate, he assailed her heart and won it. To give audience to the tempter is the next step to yielding up obedience to his will.[37]

One of the two principal powers in Mansoul, *Resistance,* quickly succumbs to an arrow from the army of Diabolus. The promises of Diabolus are familiar: to *enlarge the town* of man-soul, *to augment their freedom* and in the subtlety of pattern identical to Eden, *challenging the prohibition of the Tree of Knowledge* itself.

Dr. Maguire continues to describe this initial incursion at the Ear-Gate with the introduction of Mr. Ill-Pause, another of the diabolical army that visits Mansoul:

> Satan has many mysterious angels who are ready to second their master's temptations and to commend his wily overtures. Thus Ill-Pause persuades the men of Mansoul; and, lo! to the temptation from without (which was utterly powerless in itself), there answers the yielding from within. This is the fatal act; and is straightaway followed by another grave disaster—the death of Innocency, one of the chiefest and most honorable townsmen. His sensitive soul was poisoned by the contact of the breath of the lost.[380]

The Holy War continues with civil war raging within Mansoul and the defeat of the giant Diabolus and his demonic army by the son of Shaddai, Emmanuel, but the allegory perhaps finds more direct application in the twenty-first century than earlier. This is because, more than any other time in history, the seduction of high-tech has taken firm root—and among the most vulnerable of the population. Culturally adrift, this high risk generation, most of whom have never heard the exquisite truths of John Bunyan, has been denied spiritual cultivation through an educational system which values tolerance above absolutes and through social training that elevates technology above heritage.

Full-Fledged Ear-Gate Assault

With the advent of cell phones, iPods, and other "personal devices," the ear-gates of an entire generation have been dangerously compromised. In addition to the obvious physical risks that associate cell phone use and texting while driving, effects have been measured on teenage language abilities and a markedly increased incidence of tinnitus, a chronic "ringing-in-the-ears."

A 2005 ChildWise study found that one in four children under the age of eight had a mobile phone, a figure which increased to 89 percent by the time the child reaches eleven years.

"Teenagers: A Generation Unplugged" is a 2008 study which determined that four out of five teens carry a wireless device (an increase of 40 percent from 2004) and found that their cell phones rank second, only to clothing, in communicating personal social status and popularity, "outranking jewelry, watches and shoes." Additionally, over half (52 percent) view cell phones as a form of entertainment and 80 percent feel that a cell phone provides a sense of security while 36 percent dislike the idea of others knowing their exact location.

While a recent WHO [World Health Organization] study determined that a cell phone-cancer link is inconclusive, the UN [United Nations] did acknowledge that the 2010 examination of thirteen thousand participants found up to 40 percent higher incidence of glioma, a cancerous brain tumor, among the 10 percent that used the mobile phone most. While there is near-unanimous agreement within the scientific community that it is simply too early to accurately project damage caused by radiation, even the most modest estimates acknowledge minimal consequences, the estimated 4.6 billion cell phone users "appear prepared to take the risk" without "firm assurances" that they are safe.[381]

As dire as these incidences for physical damage appear, the psychological and spiritual implications are the significantly more profound—and sinister.

"Thought reading" has come of age. First published in January 2009, CBS revealed technology conducted at Carnegie Mellon University that makes it possible to see what is happening within the brain while people are thinking. Using specialized magnetic resonance, neuro-activity can be recorded by analyzing brain activity.[382]

While mainstream media carefully smudges the science fiction-actual science line, both government and private research groups charge the fields of neuro-fingerprinting, neuro-databases, and abject control neuro-control.

Following the Human Genome Project's mapping of human DNA, the Human Brain Project, HPB, was launched. The international research group hopes to provide a "blueprint of normal brain activity" to the goal of understanding brain function for improved health care, but inherent in the study is the very real possibility of threatening autonomous and unrestricted thought. If in 2002 the BBC was touting wireless sensors that record and generate brain waves and anatomical functions remotely,[383] and in 2008, *Scientific American* reported that scientists can "selectively control brain function by transcranial magnetic stimulation (TMS)" via the pulsing of powerful electromagnetic fields into the brain or a subject's brain circuits,[384] what might be a more current—and sinister—application?

Unbounded Evil

A March 2010 study published in the *Proceedings of the National Academy of Sciences* reported that electromagnetic currents directed at the right temperoparieto junction (TPJ), located just above and behind the right ear (the same location mentioned above from Exodus 29:20 where the priests were to be anointed that they might hear from God), can impair a person's ability to make moral judgments by inducing a current which disrupts this region of the brain.

By producing "striking evidence" that the right TPJ is "critical for making moral judgments," the lead author, Liane Young, also noted that "under normal circumstances, people are very confident and consistent in these kinds of moral judgments." The researchers believe that transmagnetic stimulation, TMS, interfered with the subject's ability to interpret the intentions of others, suggesting that they are believed to be "morally blame-worthy."[385] Subsequent publications have proposed an interest by the U.S. military to use transmagnetic stimulation to enhance soldiers' battle duration by reducing the need to stop for sleep.[386]

With the acknowledged identification, documentation and cataloging of "brain-printing" via wireless devices, and the comparatively recent release of the morally consequential findings of transmagnetic stimulation, the premise of Stephen King's 2006 novel, *Cell,* evokes a frighteningly possible scenario:

> Mobile phones deliver the apocalypse to millions of unsuspecting humans by wiping their brains of any humanity, leaving only aggressive and destructive impulses behind....
>
> What if cell phones didn't cause cancer? What if they did something much worse? What if they turned the user into a zombie killing machine?[387]

Or perhaps just a glance at a keyboard before powering down: The message is clear: CONTROL ALTER DELETE

Mind Gates—From Nightmares to Inception

From Sue Bradley's chilling e-mail above discussing how the area of the right ear (which was to be anointed for priestly hearing of God in the Old Testament) is now being targeted by electromagnetic currents to illustrate how a person's moral judgment could be impaired, to the work of neurologist Olaf Blanke that produced a "shadow person" by stimulating the left TPJ at the left ear, serious questions arrive about the mysteries of the mind and what God may know that we don't (and therefore why the priests were anointed there) about spiritual gateways existing in these regions. Once again, by interfacing with or manipulating the brain in this way, are we approaching a forbidden unknown?

Another example of how near-horizon neurosciences and human-machine integration may reconfigure human brains to allow borderline (or more than borderline) supernatural activity involves certain video games played before bedtime, which are being shown to allow people to take control of their dreams, to shape the alternate reality of dream worlds in a way that reflects spiritual warfare. According to *LiveScience* senior writer Jeremy Hsu, published studies on the dreams of hard-core gamers by Jayne Gackenbach, a psychologist at Grant MacEwan University in Canada, found that gamers experienced reversed-threat simulation in nightmares, which allowed the dreamer *to become the threatener instead of the threatened.* In other words, a scary nightmare scenario turned into something "fun" for a gamer, allowing the player to assume the role of the aggressor or demon attacker. "They don't run away; they turn and fight back. They're more aggressive than the norms," Gackenbach explained. "Levels of aggression in gamer dreams also included hyper-violence not unlike that of an R-rated movie," and when these dreaming gamers became aggressive, "oh boy, they go off the top."[388]

From learning to influence our private dreams via game-tech to having our dreams infiltrated and manipulated by outside forces, disquieting ideas deepen. In the 2010 movie *Inception* starring Leonardo DiCaprio, industrial spies use a dream machine called PASIV to steal corporate secrets by means of invasion and "extraction" of private information through a victim's dreams. In a second scenario, the film depicts ideas planted in the person's mind (inception) so that the individual perceives them as his or her own, thus allowing the victim to be steered toward particular decisions or actions—a modern upgrade on brainwashing a la the *Manchurian Candidate.* While the film *Inception* is fantasy, it is based in part on near-future technology. Electroencephalograms, functional magnetic resonance imaging (fMRIS), and computed tomography

(CT) scans are already being used to "read and even influence the brain," points out Aaron Saenz at the Singularity Hub. But could the fundamental science that the film *Inception* examines actually be setting the stage for making it a reality? "We're certainly working towards it," Saenz adds, continuing:

> In the next few decades we could have the means to understand, perhaps in rather detailed terms, what a person is thinking. Once that barrier is passed, we may develop the means to influence what someone thinks by directly stimulating their brain. [So] while the mind is still a very mysterious place, it may not remain that way forever.[389]

This trend toward technological mind invasion and mind control is or should be a frightening proposal for most people, especially those who value the concept of *free will*. That is because most secular neuroscientists view free will as an outdated religious notion related to "a fictional omnipotent divinity" (God) who chooses not to interfere with the choices of individuals, thus leaving them morally accountable for their actions and future judgment. There is even a concerted effort on the part of some neuroscientists to find proof against free will to illustrate that man is little more than an automaton whose decisions are predetermined by a complex mixture of chemical reactions, past events, and even nature, which work together to determine one's course of action. In the 1970s, Prof. Benjamin Libet of the University of California in San Francisco claimed to have discovered proof of this theory through a series of tests in which a "time gap" between a brain's decision to act and the person's awareness of this decision led to the activity being carried out by the individual. His findings ignited a stormy debate regarding the ancient philosophical question of free will, says Naomi Darom for the online edition of *Haaretz* newspaper in Israel. "Are our decisions, the basis for our ostensible free activities, made before we are aware of them? In other words, does the brain ostensibly decide for us? And to what extent do we actually make our decisions consciously?" Prof. Hezi Yeshurun explained how those engaged in the brain research concluded "the question of free will is meaningless, because…the fact that your brain has actually decided in your absence and that I can know what you've decided before you do, paints a picture of an automaton."[390]

To insinuate that a section of the human brain makes decisions ahead of man's independent awareness of them opens a wellspring of opportunity for civil or military arms technology to target that aspect of the brain and to develop methods for "inserting" ideas in minds. DARPA, American Technology Corp., Holosonic Research Labs, and others are working on methods to adapt this science, where thoughts and ideas can be projected or "implanted" in the brain and perceived by the individual as his or her own. A while back, *Wired* magazine reported on DARPA's "sonic projector" as well as troops studying the Long Range Acoustic Device (LRAD) as a modified "Voice of God" weapon:

> It appears that some of the troops in Iraq are using "spoken" (as opposed to "screeching") LRAD to mess with enemy fighters. Islamic terrorists tend to be superstitious and, of course, very religious. LRAD can put the "word of God" into their heads. If God, in the form of a voice that only you can hear, tells you to surrender, or run away, what are you gonna do?[391]

Wired went on to acknowledge how, beyond directed sound, "it's long been known that microwaves at certain frequencies can produce an auditory effect that sounds like it's coming from within someone's head (and there's the nagging question of classified microwave work at Brooks Air Force Base that the Air Force stubbornly refuses to talk about)." It is also reported that the Pentagon tested similar research during the Gulf War of 1991 using a technology called silent sound spread spectrum (SSSS), which evidently led to the surrender of thousands of Iraqi soldiers who began "hearing voices."

People of faith, including church theologians and philosophers, should find the idea of using technology to read the minds and manipulate the thoughts of individuals indefensible, as the vanguard of free will is fundamental to our religious and philosophical ethic. To humans, autonomy of thought is the most basic of doctrines in which man is unrestrained by causality or preordained by mystical powers. Yet how these issues—neurosciences, brain-machine interfacing, cybernetics, mind control, and even free will—could actually represent a prophetic confluence of events that soon will combine in an ultimate showdown over the liberty of man may be an unavoidable and *beastly* aspect of end-times prophecy.

Notes

1. S. Jonathon O'Donnell, "Secularizing Demons: Fundamentalist Navigations in Religion and Secularity," abstract, *Zygon: Journal of Religion and Science*, issue titled *Nuclear Waste, Conspiracies, and E-Meters: Remarkable Religion and Technology* (Vol. 51, No. 3, September 2016), 640.
2. The first instance of this appears on: S. Jonathon O'Donnell, "Secularizing Demons," 642.
3. Ibid.
4. Ibid., 643.
5. Ibid.
6. Ibid., 645.
7. Ibid.
8. Ibid., 644.
9. Ibid., 645.
10. Ibid., 646.
11. Ibid., 649.
12. José M. R. Delgado, *Physical Control of the Mind: Toward a Psychocivilized Society* (Harper and Row, 1969), p. 246.
13. James Gunn, *Alternate Worlds: The Illustrated History of Science Fiction* (Prentice-Hall, 1975), p. 13.
14. John Ogilvie, *The Imperial Dictionary of the English Language: A Complete Encyclopedic Lexicon, Literary, Scientific, and Technological* (Blackie & Son, 1883, edited by Charles Annandale), p. 415.
15. Lincoln Cannon, "The Purpose of the Mormon Transhumanist Association," Mormon Transhumanist Association annual meeting, April

5, 2013, Salt Lake City, UT, as taken from my notes and media from the event. This speech can be viewed on MTA's YouTube channel: https://youtu. be/iFhQf2R_IVs.

16. Cryonics is the technique of suspending a deceased person in a frozen state until such a time when science is able to resurrect the dead individual.

17. Max More, interviewed by Carl Teichrib for Magnum Veritas Productions, LLC., June 16, 2013, GF2045 Congress. Interview location: Empire Hotel, New York City.

18. Brian Alexander, *Rapture: How Biotech Became the New Religion* (Basic Books, 2003), p. 3.

19. Hank Pellissier and Teresa Dal Santo, "Transhumanists: Who Are They, What Do They Want, Believe and Predict?" *Journal of Personal Cyberconsciousness* (Terasem Movement, Inc, 2013), Volume 8, Issue 1, p. 22–24.

20. Two studies of import: Robert M. Geraci, *Virtually Sacred: Myth and Meaning in World of Warcraft and Second Life* (Oxford University Press, 2014), and William Sims Brainbridge, *eGods: Faith Versus Fantasy in Computer Games* (Oxford University Press, 2013).

21. Pellissier and Dal Santo, "Transhumanists: Who Are They, What Do They Want, Believe and Predict?" *Journal of Personal Cyberconsciousness*, p. 22.

22. *Who Are the IEET's Audience?* Institute for Ethics and Emerging Technologies, July 16, 2013, [https://ieet.org/index.php/IEET2/more/poll20130716]; *What Do Technoprogressives Believe in 2017?* Institute for Ethics and Emerging Technologies, February 18, 2017 (https://ieet.org/index.php/IEET2/more/hughes20170218).

23. *The Shape of Things to Come* is the title of H. G. Wells' 1933 technocratic novel. "Machines of Loving Grace" is part of the techno-utopian poem, *All Watched Over by Machines of Loving Grace*, penned by American poet, Richard Brautigan, and published in 1967.

24. On Marxism, see James Steinhoff, "Transhumanism and Marxism: Philosophical Connections," *Journal of Evolution and Technology*, Volume 24, Issue 2, May 2014 (http:// jetpress.org/v24/steinhoff.pdf).

25. Stefan Lorenz Sorgner, "Nietzsche, the Overhuman, and Transhumanism," *Journal of Evolution and Technology*, Volume 20, Issue 1, March 2009 (http://jetpress.org/v20/ sorgner.pdf).

26. Eugenics was popularized in the early twentieth century throughout much of the Western world. The United States engaged in positive and negative

eugenics, as did England, Canada, Australia, Denmark, Sweden (which continued into the 1970s), and other nations—including colonial-ruled countries. The Soviet Union worked to configure a socialist eugenics program. Nazi Germany employed eugenics within its framework as "racial hygiene," and because of this, eugenics became tinctured by the horrors of the Third Reich. For more information on eugenics in narrow and broad applications, see Edwin Black, *War Against the Weak: Eugenics and America's Campaign to Create a Master Race* (Four Walls Eight Windows, 2003); Richard Weikart, *From Darwin to Hitler: Evolutionary Ethics, Eugenics, and Racism in Germany* (Palgrave Macmillan, 2004); Marc Hillel, *Of Pure Blood* (Ferni Publishers, 1979); Victoria F. Nourse, *In Reckless Hands: Skinner v. Oklahoma and the Near-Triumph of American Eugenics* (W. W. Norton & Company, 2008); Robert Proctor, *Racial Hygiene: Medicine Under the Nazis* (Harvard University Press, 1988); Stefan Kuhl, *The Nazi Connection: Eugenics, American Racism, and German National Socialism* (Oxford University Press, 1994); David Plotz, *The Genius Factory: The Curious History of the Nobel Prize Sperm Bank* (Random House, 2005); Andrew Kimbrell, *The Human Body Shop: The Cloning, Engineering, and Marketing of Life* (Regnery Publishing, 1997).

27. Dmitry Itskov, interviewed by Carl Teichrib for Magnum Veritas Productions, LLC., June 16, 2013, GF2045 Congress. Interview location: Empire Hotel, New York City.

28. Max More, interviewed by Carl Teichrib for Magnum Veritas Productions, LLC., June 16, 2013, GF2045 Congress. Interview location: Empire Hotel, New York City.

29. Anders Sandberg, interviewed by Carl Teichrib for Magnum Veritas Productions, LLC., June 16, 2013, GF2045 Congress. Interview location: Empire Hotel, New York City.

30. Quoted by Natalie Ball, "Eugenics through the Eyes of Nobel Laureates: Involvement in the Intentional Improvement of Man's Inheritable Qualities from 1905–2010," *The Proceedings of the 20th Anniversary, History of Medicine Days Conference 2011: The University of Calgary, Faculty of Medicine, Alberta, Canada* (Cambridge Scholars Publishing, 2015), p. 117.

31. See James D. Watson, *A Passion for DNA: Genes, Genomes, and Society* (Cold Springs Harbor Laboratory Press, 2000).

32. John Glad, *Future Human Evolution: Eugenics in the Twenty-First Century* (Hermitage Publishers, 2006), p. 49. Glad was controversial for his

endorsement of eugenics within the Jewish community, believing it was time for Jews to strengthen their genetic traits in order to remain a viable people group. He also boldly promoted sterilization and abortion programs for low-IQ women: "Abortion should be actively promoted, since it often serves as the last and even only resort for many low-IQ mothers who fail to practice contraception. Welfare policies need to be radically reexamined. Rather than simply pay low-IQ women more for each child, financial support should be made dependent on consent to undergo sterilization.… Eugenic family planning services are the greatest gift that the advanced countries can offer the Third World" (p. 97). For his thoughts on Jewish eugenics, see his book *Jewish Eugenics* (Wooden Shore, 2011).

33. The literature on UFO phenomena is rife with references to genetic modification. Lyssa Royal and Keith Priest in *Visitors from Within* (Royal Priest Research Press, 1992) offers channeled promises of a coming spiritual-physical upgrade in the mixing of alien and human genetics. In his book, *Abduction: Human Encounters with Aliens* (Charles Scribner's Sons, 1994), John E. Mack broached the subject of biological intervention in the abduction experience, connecting this to evolutionary processes. David M. Jacobs' troubling work, *The Threat* (Simon & Schuster, 1998), explores the topic of alien-human genetic exchange. Christian author Gary Bates explores similar themes in his volume, *Alien Intrusion: UFOs and the Evolution Connection* (Creation Book Publishers, 2004).

34. The United States Defense Advanced Research Projects Agency.

35. In April, 2017, the US Marine Corps conducted beach-landing tests to evaluate the strengths and weaknesses of an integrated nonhuman platform: robot amphibious assault vehicles, MUTT units, V-Bat drones, unmanned aerial vehicles, and other bots and connected devices. Other nations too are building up integrated robotic combat forces.

36. Lance Price, *The Modi Effect: Inside Narendra Modi's Campaign to Transform India* (Quercus, 2015), pp. 131–136.

37. D. T. Max, "Beyond Human," *National Geographic*, April 2017, Vol. 231, No. 4, p. 40ff.

38. Peter H. Diamandis, "Intelligent Self-Directed Evolution Guides Mankind's Metamorphosis into an Immortal Planetary Meta-Intelligence," *Global Future 2045 International Congress* (program), June 2013, p. 15.

39. The AI for Good Global Summit, hosted by ITU and XPRIZE—in partnership with a range of UN agencies—took place in Geneva, Switzerland, June 7–9, 2017.

40. David Gelernter, "The Second Coming—A Manifesto," *Science at the Edge: Conversations with the Leading Scientific Thinkers of Today* (Union Square Press, 2003/2008, edited by John Brockman), p. 239.

41. Gerd Leonhard, *Technology vs. Humanity: The Coming Clash Between Man and Machine* (Fast Future Publishing, 2016), p. 67.

42. Lawrence K. Frank, *Nature and Human Nature: Man's New Image of Himself* (Rutgers University Press, 1951). Frank was on the General Education Board of the Rockefeller Foundation and the National Resources Planning Board.

43. "The scientific and technological 'explosion' is transforming society.... But the big question remains: will there be enough scientists and engineers to achieve the possibilities already within technical reach?"—Ronald Schiller, "Help Wanted: Engineers and Scientists," *The Reader's Digest*, February 1967 (Canadian Edition), p. 47. For an annotative list of technical and general publications outlining scientific and engineering personnel concerns from 1965 to 1969, including workplace shortages and future industrial needs, see Frank Witham, *Scientists and Engineers in the Federal Government* (Civil Service Commission, Washington, DC, Personnel Bibliography Series Number 30, January 1970).

44. Albert Rosenfeld, "The New Man: What Will He Be Like?" *Life*, Volume 59, Number 14, October 1, 1965, p. 96.

45. Ibid., p.100. The sentence is worded awkwardly, but it makes the point.

46. Albert Rosenfeld, *The Second Genesis: The Coming Control of Life* (Prentice-Hall, 1969). Another book on accelerating technology and human change written in the same era is *The Biological Time Bomb* by Gordon Rattray Taylor (New American Library, 1968).

47. John G. Burke, "Preface," *The New Technology and Human Values* (Wadsworth Publishing Company, 1966/1968, edited by John G. Burke), p. iii.

48. Jacques Ellul, *The Technological Society* (Vintage Books, 1964).

49. C. P. Snow, *Science and Government* (Harvard University Press, 1961). Snow was involved in selecting technicians and scientific personnel to assist in England's military response to Nazi Germany.

50. Robert M. Hutchins, "Science, Scientists, and Politics," *The New Technology and Human Values* (Wadsworth Publishing Company, 1966/1968, edited by John G. Burke), pp.40–43.

51. Herbert J. Muller, *The Children of Frankenstein: A Primer on Modern Technology and Human Values* (Indiana University Press, 1970), p. 370.

52. Delgado, *Physical Control of the Mind*, pp. 221–222.

53. Ibid., p. 253.

54. 9th Colloquium on the Law of the Futuristic Persons, December 10, 2013, Terasem Island, Second Life.

55. Heikki Laakko, "Transhumanism Today and Tomorrow: Ethical Aspects and Laws Shaping Our Future," presentation at the 9th Colloquium on the Law of the Futuristic Persons, December 10, 2013, Terasem Island, Second Life. As taken from my notes of the event: Quotes are directly attributed to Laakko, while the questions are deduced from his talk. Laakko is a BCI specialist.

56. Neil Postman, *Technopoly: The Surrender of Culture to Technology* (Vintage Books, 1992/1993), p. 5.

57. Douglas Groothuis, *The Soul in Cyber-Space* (BakerBooks, 1997), p. 143.

58. Jaron Lanier, *You Are Not A Gadget: A Manifesto* (Alfred A. Knopf, 2010).

59. Ibid., p. 29.

60. Gelernter, "The Second Coming—A Manifesto," *Science at the Edge*, p. 239, italics in original.

61. Hank Pellissier and Teresa Dal Santo, "Transhumanists: Who Are They, What Do They Want, Believe and Predict?" *Journal of Personal Cyberconsciousness* (Terasem Movement, Inc, 2013), Volume 8, Issue 1, pp. 22–24.

62. *Who are the IEET's Audience?* Institute for Ethics and Emerging Technologies, July 16, 2013, (https://ieet.org/index.php/IEET2/more/poll20130716).

63. As I write this, Zoltan Istvan is campaigning for California's 2018 governor's election.

64. Zoltan Istvan, "Transhumanism, Religion, and Atheism," given at the Religion and Transhumanism Conference, May 10, 2014, Piedmont, California, hosted by the Brighter Brains Institute. His presentation was recorded and is available on the YouTube channel of the Mormon Transhumanist Association (https://youtu.be/jXOXBYr2Z7I).

65. Istvan's message has come through media interviews and conference talks, his articles and essays, and in his novel *The Transhumanist Wager* (Futurity Imagine Media, 2013).

66. Julian Huxley, *Religion Without Revelation* (Watts and Company, 1941, originally published in 1927).

67. Julian Huxley, *New Bottles for New Wine* (Harper and Brothers Publishers, 1957), p. 103.

68. Ibid., pp. 103–104.

69. Ibid., p. 251.

70. Ibid., p. 122.

71. Ibid., p. 260.

72. Ibid., pp. 279–280.

73. Ibid., p. 13.

74. Ibid., p. 17.

75. Pierre Teilhard de Chardin, *The Phenomenon of Man* (HarperPerennial, 1976, originally published in 1955), p. 221.

76. Ibid., p. 253.

77. Ibid., p. 281.

78. Ibid., p. 285, italics in original.

79. Pierre Teilhard de Chardin, *The Future of Man* (Harper Colophon Books, 1969), p. 275.

80. Pierre Teilhard de Chardin, *Christianity and Evolution* (Harcourt, Inc., 1969), p. 240, italics in original.

81. Ibid., p. 239.

82. Frank J. Tipler, *The Psychics of Immortality: Modern Cosmology, God and the Resurrection of the Dead* (Doubleday, 1994), p. 154.

83. Zoltan Istvan, "I visited one of the largest megachurches in the US as an atheist Transhumanist presidential candidate—here's what happened," *Business Insider*, Tech News section, December 2, 2015 (www. businessinsider.com/transhumanist-zoltan-istvan-visits- one-of-the-largest-megachurches-in-the-us-2015-11).

84. Zoltan Istvan, "Transhumanism," presentation at the 10th Colloquium on the Law of the Futuristic Persons, December 10, 2015, Terasem Island, Second Life. Note: In what could be viewed as a campaign stunt, he also visited Alabama's Church of the Highlands, the largest evangelical congregation in the state, where he toured the main campus and talked to one of the pastors. That is, until someone in the church did an Internet search on transhumanism, and then—for better or worse—Team Istvan was escorted from the grounds. See Zoltan Istvan, "Forget Trump, Zoltan Istvan wants to be the 'anti-death' president," *Wired*, UK online edition, November 8, 2016 (www.wired.co.uk/article/the-transhumanist-age).

85. Timothy Leary, *Design for Dying* (HarperEdge, 1997), pp. 36–37.

86. Terence McKenna, *The Archaic Revival: Speculations on Psychedelic Mushrooms, the Amazon, Virtual Reality, UFOs, Evolution, Shamanism, the*

Rebirth of the Goddess, and the End of History (HarperOne, 1991/1992), p. 19.

87. Terence McKenna, a talk given at the Esalen Institute, Big Sur, California, August 1996. An audio recording of his presentation is archived with the author. This talk, now titled *The Evolutionary Importance of Technology*, can be accessed via YouTube at https://youtu.be/0O93SEOWjE4. Using the May–October 1996 *Esalen Catalog*, I have correlated McKenna's workshop to the weekend of August 2–4, 1996.

88. Britt Welin and Ken Adams workshop, "Machine Dreams and Technoshamanism," *The Esalen Catalog*, May–October, 1993 (The Esalen Institute, 1993), p. 39.

89. Douglas Rushkoff, "Technoshamanism: Total Immersion in Spiritual Technology," *The Esalen Catalog*, September 1995–February 1996 (The Esalen Institute, 1995), p. 55.

90. Virtual Reality Markup Language is a standard file describing 3-D imaging for virtual reality applications in the World Wide Web. Today the word "Modeling" replaces "Markup."

91. Mark Pesce is well known as a techno-pagan, interacting with cybernetics through a decidedly pagan lens. See Eric Davis, "Technopagans: May the Astral Plane Be Reborn in Cyberspace," *Wired*, online edition, July 1, 1995 (www.wired.com/1995/07/technopagans). See also Eric Davis, *TechGnosis: Myth, Magic and Mysticism in the Age of Information* (Harmony Books, 1998), pp. 192–193. Davis, a contemporary of Pesce and a fellow traveler with McKenna, is himself a psychonaut and visionary futurist who lectures at Esalen and is well known within the Burning Man community.

92. McKenna, *The Archaic Revival*, p. 165.

93. Mark Pesce, *Becoming Transhuman*, a presentation given at Mindstates, Berkeley, California, May 2001; transcript of talk, p. 10.

94. Genesis 1–3.

95. Romans 5:12.

96. Isaiah 64:6, Lexham English Bible (Logos Bible Software, 2011).

97. Psalm 14:2–3.

98. Acts 4:12.

99. David F. Noble, *The Religion of Technology: The Divinity of Man and the Spirit of Invention* (Penguin Books, 1997/1999), p. 12.

100. Israel Regardie, *The Original Account of the Teachings, Rites and Ceremonies*

of the Hermetic Order of the Golden Dawn (Llewellyn Publishing, 2003, first published in 1937), pp. 10, 135.

101. W. L. Wilmshurst, *The Meaning of Masonry* (Gramercy Books, 1980, originally published in 1922), p. 47.

102. Henry C. Clausen, *Emergence of the Mystical* (Ancient and Accepted Scottish Rite of Freemasonry, Southern Jurisdiction, 1981), p. 92.

103. C. W. Leadbeater, *The Masters and the Path* (The Theosophical Press, 1925), p. 3. Other names ascribed to advanced soul beings include, "Great Ones, the Planetary Spirits, Great Angels, Karmic Deities, Dhyan Chohans, Buddhas, Christs and Masters." (p. 200).

104. Ibid., p. 1.

105. The concept of the Universal Over-Soul plays out in Terence McKenna's idea of the Overmind. It is visible in the Rosicrucian and esoteric model of the Cosmic Christ, and in the New Age notion of Cosmic Consciousness. Each speaks to the same value: All is One.

106. H. A. W. Coryn, "Consciousness," *Lucifer*, Vol. 9, No. 50, October 15, 1891, p. 125.

107. P. D. Ouspensky, *Tertium Organum: The Third Canon of Thought, a Key to the Enigmas of the World* (Alfred A. Knopf, 1922, second edition, originally published in Russian in 1912), p. 318.

108. Ray Kurzweil, *The Singularity Is Near: When Humans Transcend Biology* (Penguin Books, 2005), pp. 388–389.

109. Zoltan Istvan, "Transhumanism," presentation at the 10th Colloquium on the Law of the Futuristic Persons, December 10, 2015, Terasem Island, Second Life. Recording of the event on file.

110. Psalm 139:14, "I will praise You, for I am fearfully and wonderfully made; Marvelous are Your works, and that my soul knows very well."

111. See chapter 11.

112. 2 Thessalonians 2:1–12.

113. Noble, *The Religion of Technology*, p. 121.

114. George M. Young, *The Russian Cosmists: The Esoteric Futurism of Nikolai Fedorov and his Followers* (Oxford University Press, 2012), p. 3.

115. Ibid., p. 25.

116. Ibid., pp. 145–154.

117. Marina Benjamin, *Rocket Dreams: How the Space Age Shaped Our Vision of a World Beyond* (Free Press, 2003), p. 122. See also Young, *The Russian Cosmists*, p. 148–149 and Noble, *The Religion of Technology*, p. 120–121.

118. Ibid., p. 123. Tsiolkovsky also believed in self-perfection by weeding the human population of defective individuals. See Young, *The Russian Cosmists*, pp. 151–152.

119. Young, *The Russian Cosmists*, p. 156.

120. Ibid., p. 157.

121. Ibid., p. 158.

122. Dmitry Itskov, interviewed by Carl Teichrib for Magnum Veritas Productions, LLC., June 16, 2013, GF2045. Interview location: Empire Hotel, New York City.

123. *2045: A New Era for Humanity* (2045 Initiative, released in 2012 as a lead-up to GF2045). This video is available online.

124. The World Transhumanist Association later morphed into Humanity+, an influential transhuman think tank.

125. The quote comes from a sidebar paragraph in Giulio Prisco's Turing Church blog (http://turingchurch.com), accessed June 2, 2017.

126. As taken from his online CV (http://goertzel.org/Goertzel_resume.pdf).

127. Ben Goertzel, *A Cosmist Manifesto: Practical Philosophy for the Posthuman Age* (Humanity+ Press, 2010), pp. 235–236.

128. Ibid., p. 236.

129. Gabriel Rothblatt, speech titled *Terasem Movement Transreligion*, given during the Mormon Transhumanist Association annual conference, April 5, 2013. A video copy of his presentation is posted on the MTA YouTube channel (https://youtu.be/BcQqFYXeFDo).

130. Martine Rothblatt, *Virtually Human: The Promise—and the Peril—of Digital Immortality* (St. Martin's Press/Picador, 2014), p. 270.

131. Ibid., pp. 268–269.

132. Martine Rothblatt, *The Apartheid of Sex: A Manifesto on the Freedom of Gender* (Crown, 1995).

133. Martine Rothblatt, *From Transgender to Transhuman: A Manifesto on the Freedom of Form* (Martine Rothblatt, edited by Nickolas Mayer, 2011), p. xiv, italics in original.

134. Lincoln Cannon, then president of the MTA, made the statement during the 2013 MTA conference (April 5). For more on the connection between Mormonism and transhumanism, see Allsop et al, "Complementary Aspects of Mormonism and Transhumanism," *Parallels and Convergences: Mormon Thought and Engineering Visions* (Greg Kofford Books, 2012, edited by A. Scott Howe and Richard L. Bushman), pp. 67–92.

135. *Mormon Transhumanist Member Survey 2014 Summary*, on file.

136. From what I have gathered, most have failed due to internal tensions and clashes.

137. The MTA is mentioned in the following works: Robert M. Geraci, *Apocalyptic AI: Visions of Heaven in Robotics, Artificial Intelligence, and Virtual Reality* (Oxford University Press, 2010); Gregory R. Hansell and William Grassie (editors), *H+/-: Transhumanism and Its Critics* (Metanexus Institute, 2011); Catherine Wessinger (editor), *The Oxford Handbook of Millennialism* (Oxford University Press, 2011); Maxwell J. Mehlman, *Transhumanist Dreams and Dystopian Nightmares: The Promise and Peril of Genetic Engineering* (John Hopkins University Press, 2012); Morgan Luck (editor), *Philosophical Explorations of New and Alternative Religious Movements* (Routledge, 2012); Calvin Mercer and Derek F. Maher (editors), *Transhumanism and the Body: The World Religions Speak* (Palgrave Macmillan, 2014); William Sims Bainbridge, *Dynamic Secularization: Information Technology and the Tension Between Religion and Science* (Springer International Publishing, 2017).

138. A. Scott Howe and Richard L. Bushman (editors), *Parallels and Convergences: Mormon Thought and Engineering Vision* (Greg Kofford Books, 2012).

139. Quoted by Carl Teichrib, "The Rise of Techno-Gods: The Merging of Transhumanism and Spirituality," *Forcing Change*, Volume 4, Issue 10, October 2010, p. 10–11, as drawn from audio recordings and research notes from the conference.

140. Ibid., p. 11.

141. Carl Teichrib, *A Conservative Christian Critique of Religious Transhumanism*, speech given to the Mormon Transhumanist Association, April 5, 2013. Posted on the MTA YouTube channel: https://youtu.be/VFxDVZjyjeE. You can also read a copy of the speech in *Forcing Change*, Volume 7, Issue 4, April 2013, pp. 14–18.

142. Luke 23:39–43.

143. A great example from the Old Testament is King David in Psalm 51. "For You do not desire sacrifice, or else I would give it; You do not delight in burnt offering. The sacrifices of God are a broken spirit, a broken and a contrite heart—these, O God, You will not despise" (Psalm 51:16–17).

144. Consider who raised Jesus from the dead: Jesus Himself—"No one takes it from Me, but I lay it down of Myself. I have power to lay it down, and

I have power to take it again. This command I have received from My Father" (John 10:18); God the Father—"whom God raised up, having loosed the pains of death, because it was not possible that He should be held by it" (Acts 2:24); And the Spirit—"But if the Spirit of Him who raised Jesus from the dead dwells in you, He who raised Christ from the dead will also give life to your mortal bodies through His Spirit who dwells in you" (Romans 8:11).

145. Colossians 1:18.

146. See Ephesians 4:17–23.

147. See Colossians 3.

148. See James 2:14–26.

149. See Romans 6:9.

150. Dmitry Itskov, interviewed by Carl Teichrib for Magnum Veritas Productions, LLC., June 16, 2013, GF2045. Interview location: Empire Hotel, New York City.

151. "Future Prospects of 2045 Initiative for Society," *Global Future 2045 International Congress* (official program), June 2013, p. 5.

152. Diamandis, *Global Future 2045 International Congress* (official program), June 2013, p. 15.

153. Natasha Vita-More, interviewed by Carl Teichrib for Magnum Veritas Productions, LLC., June 16, 2013, GF 2045. Interview location: Empire Hotel, New York City.

154. Alan Harrington, *The Immortalist* (Celestial Arts, 1969/1977), p. 3.

155. Tipler, *The Psychics of Immortality*, p. 17.

156. Ray Kurzweil, *The Age of Spiritual Machines: When Computers Exceed Human Intelligence* (Penguin Books, 1999/2000), pp. 128–129, italics in original.

157. Kurzweil, *The Singularity is Near*, p. 325.

158. Giulio Pisco's mini-manifesto has been reprinted in Ben Goertzel's book, *A Cosmist Manifesto* (Humanity+ Press, 2010), pp. 10–11.

159. Zoltan Istvan, "Transhumanism," presentation at the 10th Colloquium on the Law of the Futuristic Persons, December 10, 2015, Terasem Island, Second Life.

160. The first RAAD Fest took place August 4–7, 2016 and August 9–13, 2017.

161. Harrington, *The Immortalist*, p. 273.

162. Ibid., p. 273.

163. Richard L. Bushman, "Forward," *Parallels and Convergences*, p. vi.

164. Lincoln Cannon, *The Purpose of the Mormon Transhumanist Association* (presentation), Mormon Transhumanist Association annual meeting, April 5, 2013, Salt Lake City, UT.

165. Richard Seed,, *Technocalyps*, written and directed by Frank Theys (Votnik, 2006).

166. Ibid.

167. Kevin Kelly, *Out of Control: The Rise of Neo-Biological Civilization* (Addison-Wesley Publishing, 1994), p. 233. The book was also published under the title, *Out of Control: The New Biology of Machines, Social Systems, and the Economic World*.

168. Ibid., p. 4.

169. Dr. John Glad, interviewed by Carl Teichrib for Magnum Veritas Productions, LLC., February 19, 2013. Location: John Glad's apartment, Washington, DC.

170. Glad, *Future Human Evolution*, p. 104–105.

171. Goertzel, *A Cosmist Manifesto*, p. 29, italics in original.

172. Kurzweil, *The Singularity Is Near*, p. 390.

173. Pesce, *Becoming Transhuman*, transcript p. 12.

174. Ben Goertzel's *A Cosmist Manifesto* is dedicated to Valentin Turchin, quoting Turchin's closing paragraph from *The Phenomenon of Science*; italics in Turchin's original text.

175. I interviewed James Martin at GF2045 on June 16, 2013. After the interview, Dr. Martin and I walked from the Empire Hotel back to the Lincoln Center, and during our chat, I discovered that he, like myself, grew up on a farm. This commonality immediately created a sense of connection, and he invited me for a meal and a chance to talk off the record. It was a fantastic opportunity, but unfortunately I had to decline as another interview was scheduled in the next few minutes. He expressed interest in a future meeting and we reluctantly parted. But I never talked to him again. A few days later, on June 24, a kayaker found Dr. Martin dead in the Bermuda Sea not far from his private island. Upon hearing the news of his passing, it felt like a fifty-pound weight had hit me. The sobering yet enduring words of Isaiah 40 rings true: "All flesh is grass, and all its loveliness is like the flower of the field. The grass withers, the flower fades, but the word of our God stands forever."

176. Solar-powered cities in the far north and distant south infer tie-in to a

global energy grid, for the shortened sunlight hours in winter would drastically reduce solar production during months when energy use would be the highest and most critical.

177. Dmitry Itskov, "On the Path to a New Evolutionary Strategy," *Global Future 2045 International Congress* (official program), June 2013, p. 1.

178. As taken from my notes of the event.

179. Ibid.

180. Ibid.

181. Ecclesiastes 1:9.

182. Genesis 3:4–5.

183. Postman, *Technopoly*, p. 165.

184. Michael Garfield, "The Psychedelic Transhumanists," Transhumanist Plus, September 29, 2009, (http://hplusmagazine.com/2009/09/29/psychedelic-transhumanists).

185. Congressional Record, vol. 118, No. 26 (1974).

186. John Horgan, "Tribute to Jose Delgado, Legendary and Slightly Scary Pioneer of Mind Control," *Scientific American*, September 25, 2017, (https://blogs.scientificamerican.com/cross-check/tribute-to-jose-delgado-legendary-and-slightly-scary-pioneer-of-mind-control).

187. "God did not spare angels when they sinned, but cast them into hell and committed them to chains of gloomy darkness to be kept until the judgment" (2 Peter 2:4, ESV). The only place in the Bible where we are told of angels sinning is Genesis 6:1–4 (the Watchers/*apkallu,* who took human wives and produced the Nephilim). The Greek verb translated "cast them into hell" is *tartaroo,* which literally means "thrust down to Tartarus." In Greek cosmology, Tartarus was a special place of punishment located as far below Hades (Hell) as Earth was below Heaven. Since Peter wrote under the guidance of the Holy Spirit, we assume he knew the difference between Tartarus and Hades. It is the only place in the Bible where that word is used.

188. "Enmerkar and the Lord of Aratta: translation." *The Electronic Corpus of Sumerian Literature,* (http://etcsl.orinst.ox.ac.uk/section1/tr1823.htm), retrieved 12/27/17.

189. Becker, Helmut and Fassbinder, Jörg W. E. (2003), "Magnetometry at Uruk (Iraq): The City of King Gilgamesh," *Archaeologia Polona, 41*, pp. 122–124.

190. 1 Enoch 6:6. Although the scholar Edward Lipinski suggested in his 1971 paper "El's Abode" that "days of Jared" should read "days of the *yarid,*"

which was a ritual libation—a drink offering for the gods. As Lipinski noted, the summit of Mount Hermon is scooped out, and earlier scholars, such as Charles Clermont-Ganneau in 1903, speculated that this may have been where worshipers poured their liquid offerings.

191. Nickelsburg, George W. E. *1 Enoch: The Hermeneia Translation.* Fortress Press. Kindle Edition, p. 26.

192. Greenfield, J. C. (1999). "Apkallu," *Dictionary of Deities and Demons in the Bible.* Van der Toorn, K., Becking, B., & Van der Horst, P. W. (Eds.). Brill, p. 73.

193. Annus, Amar (2010). "On the Origin of Watchers: A Comparative Study of the Antediluvian Wisdom in Mesopotamian and Jewish Traditions." *Journal for the Study of the Pseudepigrapha*, Vol 19, Issue 4, pp. 277–320.

194. George, Andrew (1999). *The Epic of Gilgamesh* (London: Penguin Books), 111–112.

195. Lipi ski, Edward (1971). "El's Abode: Mythological Traditions Related to Mount Hermon and to the Mountains of Armenia," *Orientalia Lovaniensa Periodica II*, p. 19.

196. Ibid.

197. Livingston, David (2003). "Who Was Nimrod?" (http://davelivingston.com/nimrod.htm), retrieved 12/27/17.

198. "Anunna." *Ancient Mesopotamian Gods and Goddesses.* (http://oracc.museum.upenn.edu/amgg/listofdeities/anunna/index.html), retrieved 12/27/17.

199. Pritchard, James B., ed. (2010). *The Ancient Near East: An Anthology of Texts and Pictures* (Princeton University Press), p. 34.

200. Lieck, Gwendolyn (1998). *A Dictionary of Ancient Near Eastern Mythology* (New York City, New York: Routledge), p. 141.

201. George, *op. cit.*, p. 199.

202. This is well established, but see, for example: Spronk, Klaas (1986). *Beatific Afterlife in Ancient Israel and in the Ancient Near East.* Kevelaer: Butzon & Bercker, Neukirchen-Vluyn.

203. Frölich, Ida (2014). "Mesopotamian Elements and the Watchers Traditions," in *The Watchers in Jewish and Christian Tradition* (ed. Angela Kim Hawkins, Kelley Coblentz Bautch, and John Endres; Minneapolis: Fortress), p. 23.

204. Annus, Amar (2000). "Are There Greek Rephaim? On the Etymology of Greek *Meropes* and *Titanes*," *Ugarit Forschungen* 31 (1999), 13–30.

205. Jude 6.

206. Smith, Wesley J. "The Trouble with Transhumanism," *Christian Life Resources*, http://www.christianliferesources.com/article/the-trouble-with-transhumanism-1191, retrieved 12/24/17.

207. Danaylov, Nikola. "A Transhumanist Manifesto (Redux)." *Singularity Weblog*, March 11, 2016. (https://www.singularityweblog.com/a-transhumanist-manifesto/), retrieved 12/24/17.

208. Ibid.

209. (http://www.zoltanistvan.com/TranshumanistWager.html), retrieved 12/24/17.

210. John 15:13.

211. Istvan, Zoltan. *The Transhumanist Wager*. Futurity Imagine Media LLC, 2013, pp. 127–128.

212. Searle, Rick. "Betting Against the Transhumanist Wager." Institute for Ethics and Emerging Technologies, September 16, 2013. https://ieet.org/index.php/IEET2/more/searle20130916, retrieved 12/25/17.

213. Which is just one of several key points of LDS doctrine that disqualifies Mormonism as a Christian denomination.

214. (https://www.christiantranshumanism.org), retrieved 12/28/17.

215. 1 Corinthians 15:3–4, ESV.

216. 1 John 3:8.

217. See Matthew 16:13–17:13. Jesus led Peter, James, and John "up a high mountain" in the vicinity of Caesarea Philippi, which can only be Hermon.

218. Kumparak, Greg (2014). "Elon Musk Compares Building Artificial Intelligence to 'Summoning the Demon'," *TechCrunch*, October 26, 2014. (https://techcrunch.com/2014/10/26/elon-musk-compares-building-artificial-intelligence-to-summoning-the-demon/), retrieved 12/31/17.

219. Benek, Christopher J. "How to Prevent an Artificial Intelligence God," *The Christian Post*, October 20, 2017. (https://www.christianpost.com/news/how-to-prevent-an-artificial-intelligence-god-203458/), retrieved 12/28/17.

220. Ibid.

221. Sadly, most American Christians don't, either.

222. Benek, *op. cit.*

223. Genesis 1:28.

224. Matthew 28:19.

225. See 1 Corinthians 15:35–58.

226. Istvan, Zoltan. *The Transhumanist Wager.* Futurity Imagine Media LLC, 2013, pp. 127–128.

227. Murphy, Timothy F., and Marc A. Lappé (1994). *Justice and the Human Genome Project.* Berkeley: University of California Press, p. 18.

228. Ko, Lisa. "Unwanted Sterilization and Eugenics Programs in the United States," *Independent Lens*, January 29, 2016. (http://www.pbs.org/independentlens/blog/unwanted-sterilization-and-eugenics-programs-in-the-united-states/), retrieved 12/28/17.

229. Regalado, Antionio (2017). "U.S. Panel Endorses Designer Babies to Avoid Serious Disease," *MIT Technology Review*, February 14, 2017. (https://www.technologyreview.com/s/603633/us-panel-endorses-designer-babies-to-avoid-serious-disease/), retrieved 12/31/17.

230. Smith, Wesley J. "Netherlands Push to Euthanize Children," *National Review Online*, April 28, 2016. (http://www.nationalreview.com/corner/434712/netherlands-push-euthanize-children), retrieved 12/28/16.

231. For example, Vita-More, Natasha. "Transhuman: A Brief History." (http://www.natasha.cc/quiz.htm#Transhuman%20History), retrieved 12/28/17.

232. Bullinger, E. W. (1903). *The Apocalypse or "The Day of the Lord."* London: Eyre & Spottiswoode.

233. He believed that the Church Age started at Acts 28:28 rather than Pentecost, and that Paul's authoritative teaching began after the conclusion of the Book of Acts.

234. Bullinger, E. W. (1902). *The Rich Man and Lazarus or "The Intermediate State,"* London: Eyrie & Spottiswoode.

235. Schadewald, Robert J. (2000). *The Plane Truth: A History of the Flat Earth Movement.* (http://www.cantab.net/users/michael.behrend/ebooks/PlaneTruth/pages/Chapter_04.html), retrieved 12/30/17.

236. Mayo Clinic (2014). "EEG Definition," (https://www.mayoclinic.org/tests-procedures/eeg/basics/definition/prc-20014093), retrieved 12/30/17.

237. "Most American Christians Do Not Believe that Satan or the Holy Spirit Exist," (https://www.barna.com/research/most-american-christians-do-not-believe-that-satan-or-the-holy-spirit-exist/), retrieved 12/31/17.

238. Kurzweil, Raymond (2005). *The Singularity Is Near: When Humans Transcend Biology.* New York: Penguin Books, p. 7.

239. Honan, Daniel. "Ray Kurzweil: The Six Epochs of Technology Evolution," *Big Think.* (http://bigthink.com/the-nantucket-project/ray-kurzweil-the-six-epochs-of-technology-evolution), retrieved 12/30/17.

240. Draper, Lucy (2015). "Could Artificial Intelligence Kill Us Off?"
     Newsweek, June 24, 2015. (http://www.newsweek.com/artificial-
     intelligenceomega-pointai-603286), retrieved 12/30/17.

241. O'Connell, Gerard (2017). "Will Pope Francis Remove the Vatican's
     'Warning' from Teilhard de Chardin's Writings?" *America: The Jesuit
     Review*, November 21, 2017. (https://www.americamagazine.org/
     faith/2017/11/21/will-pope-francis-remove-vaticans-warning-teilhard-de-
     chardins-writings), retrieved 12/30/17.

242. Steinhart, Eric (2008). "Teilhard de Chardin and Transhumanism,"
     Journal of Evolution and Technology, Vol. 20, Issue 1, 1–22. (http://jetpress.
     org/v20/steinhart.htm), retrieved 12/31/17.

243. Ibid.

244. For example, as I write this from my living room couch, there are six
     Internet-connected devices in view, three more in the next room, and seven
     more in our office on the other side of the wall.

245. "The Death of Gilgamesh: translation." *The Electronic Text Corpus of
     Sumerian Literature*, (http://etcsl.orinst.ox.ac.uk/section1/tr1813.htm),
     retrieved 12/31/17.

246. Marchesi, Gianni (2004). "Who Was Buried in the Royal Tombs of Ur? The
     Epigraphic and Textual Data," Orientalia, Nova Series, Vol. 73, No. 2, p. 154.

247. Ashlee Vance, "Merely Human? That's So Yesterday," *New York Times,*
     (6/11/10) (http://www.nytimes.com/2010/06/13/business/13sing.
     html?_r=1).

248. Wesley J. Smith, "Pitching the New Transhumanism
     Religion in the NYT," *First Things* (6/14/10) (http://www.
     firstthings.com/blogs/secondhandsmoke/2010/06/14/
     pitching-the-new-transhumanism-religion-in-the-nyt/).

249. Gregory Jordan, "Apologia for Transhumanist Religion," *Journal of
     Evolution and Technology, Published by the Institute for Ethics and Emerging
     Technologies* (2005) (http://jetpress.org/volume15/jordan2.html).

250. Brent Waters, "The Future of the Human Species (Part 1)," (http://www.
     cbhd.org/content/future-human-species).

251. As quoted by C. Christopher Hook in "The Techno Sapiens Are Coming,"
     Christianity Today (January 2004) (http://www.christianitytoday.com/
     ct/2004/january/1.36.html).

252. Ibid.

253. Thomas and Nita Horn, *Forbidden Gates*, Defender Publishing, 2011, 252–257.

254. (http://marvel.com/universe/Spider-Man_(Peter_Parker)#axzz514Qb1W00), accessed December 12, 2017.

255. (http://www.dccomics.com/characters/the-flash), accessed December 12, 2017.

256. (http://marvel.com/universe/Fantastic_Four#axzz514Qb1W00), accessed December 12, 2017.

257. (http://marvel.com/universe/Captain_America_(Steve_Rogers)#axzz514Qb1W00), accessed December 12, 2017.

258. (http://www.dccomics.com/characters/superman), accessed December 12, 2017.

259. (http://www.dccomics.com/characters/wonder-woman), accessed December 12, 2017.

260. (http://www.dccomics.com/characters/aquaman), accessed December 12, 2017.

261. (http://www.dccomics.com/characters/bizarro), accessed December 15, 2017.

262. (http://www.dccomics.com/characters/brainiac), accessed December 15, 2017.

263. (http://www.dccomics.com/characters/sinestro), accessed December 15, 2017.

264. (https://en.wikipedia.org/wiki/Giganta), accessed December 15, 2017.

265. (http://www.dccomics.com/characters/solomon-grundy), accessed December 15, 2017.

266. Bill Hamon, *The Eternal Church* (Shippensburg, PA: Destiny Image, 1981, revised 2003) 385. (http://reachouttrust.org/manifest-sons-of-god/), accessed December 15, 2017.

267. R. C. Sproul, *How Could Jesus Be Both Divine and Human?* (http://www.ntslibrary.com/Online-Library-How-Could-Jesus-Be-Both-Divine-and-Human.htm), accessed December 4, 2017.

268. (https://www.christiantranshumanism.org/).

269. *The Journal of Medicine and Philosophy: A Forum for Bioethics and Philosophy of Medicine*, Volume 35, Issue 6, 1 December 2010, Pages 656–669, (https://doi.org/10.1093/jmp/jhq052).

270. *The Definition of Morality* (https://plato.stanford.edu/entries/morality-definition/)

271. (https://en.oxforddictionaries.com/definition/human_being).

272. (https://www.globalresearch.ca/transhumanism-genetic-manipulation-and-the-age-of-human-destruction-towards-the-end-of-homo-sapiens/5373858).

273. (http://www.dictionary.com/browse/genocide).

274. www.ZoltanIstvan.com.

275. (https://www.salon.com/2016/02/19/now_this_is_an_outsider_candidate_zoltan_istvan_a_transhumanist_running_for_president_is_promoting_the_facilitation_of_immortality/).

276. (http://www.ibtimes.co.uk/zoltan-istvan-immortality-bus-delivers-transhumanist-bill-rights-us-capitol-1534388).

277. (https://www.cancer.gov/news-events/cancer-currents-blog/2017/crispr-immunotherapy).

278. (https://bible.org/article/gods-plan-salvation).

279. Don Galeon and Christianna Reedy, "Kurzweil Claims That the Singularity Will Happen by 2045," *Futurism*, October 5, 2017, (https://futurism.com/kurzweil-claims-that-the-singularity-will-happen-by-2045).

280. Tim LaHaye and Ed Hindson, *The Popular Encyclopedia of Bible Prophecy* (Eugene, OR: Harvest House Publishers, 2004), 44–45.

281. Derek P. Gilbert and Josh Peck, *The Day the Earth Stands Still* (Crane, MO: Defender, 2017), 1–2.

282. Federal Bureau of Investigation, "Majestic12," https://vault.fbi.gov/Majestic%2012.

283. Aaron Mak, "How Conspiracy Theorists on Reddit Are Reacting to That New York Times UFO Story," Slate, December 19, 2017, (http://www.slate.com/blogs/future_tense/2017/12/19/how_conspiracy_theorists_on_reddit_4chan_are_reacting_to_the_new_york_times.html).

284. Caroline Iggulden, "STRANGER THINGS HAPPENED From mind control, brainwashing and monsters—theories claim Stranger Things happened in REAL LIFE in a secret government project," *The Sun*, October 21, 2017, (https://www.thesun.co.uk/fabulous/4734230/theories-stranger-things-happened-real-life).

285. Dr. Kevin Clarkson, "Kevin's Corner," *Prophecy in the News*, January 2018, 8.

286. Camila Domonoske, "Elon Musk Warns Governors: Artificial Intelligence Poses 'Existential Risk,'" NPR, July 17, 2017, (https://www.npr.org/sections/thetwo-way/2017/07/17/537686649/elon-musk-warns-governors-artificial-intelligence-poses-existential-risk).

287. Maureen Dowd, "Elon Musk's Billion-Dollar Crusade to Stop the A.I. Apocalypse," *Vanity Fair*, March 26, 2017.

288. Mark Harris, "Inside the First Church of Artificial Intelligence," *Wired*, November 15, 2017, (https://www.wired.com/story/ anthony-levandowski-artificial-intelligence-religion).

289. Bibliography for this Chapter
North, Gary. *Unholy Spirits: Occultism and New Age Humanism*. Fort Worth, Texas: Dominion Press,1986.
Hoffman, Jordan. "Man of Steel No Longer Man of S'htell?" *Times of Israel*. 14 June 2013.
Skelton, Stephen. *The Gospel According to the World's Greatest Superhero*. Eugene, Oregon: Harvest House Publishers, 2006.

290. (http://www.cultura.va/content/dam/cultura/docs/pdf/events/ PlenaryTheme2017_en.pdf).

291. Ibid.

292. (https://www.forbes.com/sites/johnfarrell/2017/11/24/vatican-council-asks-the-pope-to-exonerate-jesuit-scientists-writings/#4f2cce5c45e8).

293. Richard Harter, "Piltdown Man: The Bogus Bones Caper," *Talk Origins* (1996), last accessed February 12, 2013, (http://www.talkorigins.org/faqs/ piltdown.html).

294. Pierre Teilhard de Chardin, *Christianity and Evolution* (New York, NY: A Harvest Book, 1971), 232.

295. Ibid., 234.

296. Malachi Martin, *The Jesuits* (New York, NY: Simon & Schuster, 1988), 290.

297. Pope Benedict XVI, *Credo for Today: What Christians Believe* (San Francisco: Ignatius Press, 2009), 34.

298. Ibid., 113.

299. Joseph Cardinal Ratzinger, *Principles of Catholic Theology: Building Stones for a Fundamental Theology*, translated by Sister Mary Frances McCarthy, S.n.d. (San Francisco: Ignatius Press, 1987), 334.

300. "Teilhard de Chardin, Pierre," *Encyclopedia of Paleontology*, s.v, last accessed January 26, 2013, (http://www.liberty.edu:2048/login?url=http://www. credoreference.com/entry/routpaleont/teilhard_de_chardin_pierre).

301. Pope Benedict XVI, "Homily of July 24, 2009," in the Cathedral of Aosta, last accessed January 26, 2013, (http://www.vatican. va/holy_father/benedict_xvi/homilies/2009/documents/ hf_ben-xvi_hom_20090724_vespri-aosta_en.html).

302. Pierre Teilhard de Chardin, *The Phenomenon of Man* (New York, NY: Harperperennial, 1955), 286.

303. M. M. Ćirković, "Fermi's Paradox: The Last Challenge for Copernicanism?" *Serbian Astronomical Journal 178* (2009), 1–20.

304. Robert Emenegger, *UFOs, Past, Present, and Future* (New York, NY: Ballantine Books, 1975), 130–147.

305. Lee Speigel, "More Believe in Space Aliens than in God According to U.K. Survey," *Huffington Post*, October 18, 2012, (http://www.huffingtonpost.com/2012/10/15/alien-believers-outnumber-god_n_1968259.html), accessed 12/07/2012.

306. Kenneth J. Delano, *Many Worlds, One God*, 104.

307. Kumparak, Greg, "Elon Musk Compares Building Artificial Intelligence to Summoning the Demon," *TechCrunch*, (https://techcrunch.com/2014/10/26/elon-musk-compares-building-artificial-intelligence-to-summoning-the-demon/), accessed January 5, 2018.

308. Transcript of Rumsfeld's DoD briefing available here: (http://archive.defense.gov/Transcripts/Transcript.aspx?TranscriptID=2636), accessed January 7, 2018.

309. Barratt, James, p. 4, *Our Final Invention: Artificial Intelligence and the End of the Human Era*, Oct. 1, 2013, St. Martin's Press, New York.

310. (http://www.dailymail.co.uk/sciencetech/article-5156943/27-millennials-say-consider-dating-robot.html), accessed January 7, 2018.

311. (https://www.calvertjournal.com/news/show/8344/russian-lawmakers-vote-to-ban-pro-suicide-social-media-groups), accessed January 7, 2018.

312. (https://www.theverge.com/2013/6/27/4431274/the-electrified-brain-the-power-and-promise-of-neural-implants), accessed January 7, 2018.

313. (http://www.independent.co.uk/life-style/gadgets-and-tech/news/nosey-smurf-gumfish-and-foggybottom-the-snooping-tools-that-may-have-got-gchq-in-hot-water-9362642.html), accessed January 7, 2018.

314. (http://www.ibtimes.com/edward-snowden-reveals-secret-decryption-programs-10-things-you-need-know-about-bullrun-edgehill), accessed January 7, 2018.

315. SIGINT doc: (https://cryptome.org/2013/09/nsa-sigint-enabling-propublica-13-0905.pdf), accessed January 7, 2018.

316. (https://readwrite.com/2013/11/14/what-is-smartdust-what-is-smartdust-used-for/), accessed January 7, 2018.

317. Tom and Nita Horn, *Forbidden Gates: How Genetics, Robotics, Artificial*

Intelligence, Synthetic Biology, Nanotechnology, and Human Enhancement Herald the Dawn of Techno-Dimensional Spiritual Warfare (Crane, MO: Defender Publishing, 2010), 125–126.

318. Joe Garreau, *Radical Evolution: The Promise and Peril of Enhancing Our Minds, Our Bodies—and What It Means to Be Human* (New York: Broadway, 2005) 71–72)

319. Ray Kurzweil, *The Singularity is Near* (New York: Penguin, 2006) 7–8.

320. Abou Farman, "The Intelligent Universe," *Maison Neuve* (8/ 2/10) http://maisonneuve.org/pressroom/article/2010/aug/2/intelligent-universe/.

321. Kurzweil, 9.

322. "The Coming Technological Singularity," presented at the VISION-21 Symposium sponsored by NASA Lewis Research Center and the Ohio Aerospace Institute (3/30–31/93).

323. Tom and Nita Horn, *Forbidden Gates: How Genetics, Robotics, Artificial Intelligence, Synthetic Biology, Nanotechnology, and Human Enhancement Herald the Dawn of Techno-Dimensional Spiritual Warfare* (Crane, MO: Defender Publishing, 2010), 130–133

324. "Brain-Computer Interface Allows Person-to-Person Communication Through Power Of Thought," *ScienceDaily* (10/6/09) (http://www.sciencedaily.com/releases/2009/10/091006102637.htm).

325. Charles Ostman, *The Internet as an Organism: Emergent Human / Internet Symbiosis* (Vienna, Austria: Thirteenth European Meeting on Cybernetics and Systems Research at the University of Vienna, April 9–12, 1996) emphasis added.

326. Daron Acemoglu and Pascual Restrepo, "Robots and Jobs: Evidence from US Labor Markets," NBER Working Paper No. 23285, March 17, 2017, *National Bureau of Economic Research*, last accessed January 12, 2018, http://www.nber.org/papers/w23285. (Note that this website requires a fee to access the original report. At the time of this writing, the report is available for free through MIT's Department of Economics at the following link: https://economics.mit.edu/files/12763.)

327. Ibid.

328. Michael Chui, James Manyika, and Mehdi Miremadi, "Where Machines Could Replace Humans—and Where They Can't (Yet)," July 2016, *McKinsey Quarterly*, last accessed January 11, 2018, (https://www.mckinsey.com/business-functions/digital-mckinsey/our-insights/where-machines-could-replace-humans-and-where-they-cant-yet), emphasis added.

329. Alex Davies, "GM Will Launch Robocars without Steering Wheels Next Year," January 12, 2018, *WIRED Magazine Online*, last accessed January 16, 2018, (https://www.wired.com/story/gm-cruise-self-driving-car-launch-2019/).

330. Timothy J. Seppala, "Waymo's Driverless Taxi Service will Open to the Public Soon," November 7, 2017, *Engadget*, last accessed January 16, 2018, (https://www.engadget.com/2017/11/07/waymo-autnomous-taxi-phoenix/).

331. Ibid.

332. Michael Chui, James Manyika, and Mehdi Miremadi, "Where Machines Could Replace Humans," (https://www.mckinsey.com/business-functions/digital-mckinsey/our-insights/where-machines-could-replace-humans-and-where-they-cant-yet).

333. Ibid., emphasis added.

334. This story appears all over the Internet in relation to the search of self-awareness in robotics. As one example for reference: Fiona Macdonald, "A Robot has Just Passed a Classic Self-Awareness Test for the First Time," July 17, 2015, *Science Alert*, last accessed January 17, 2018, (https://www.sciencealert.com/a-robot-has-just-passed-a-classic-self-awareness-test-for-the-first-time).

335. Ibid.

336. Jordan Pearson, "Watch These Cute Robots Struggle to Become Self-Aware," July 16, 2015, *Motherboard*, last accessed January 17, 2018, (https://motherboard.vice.com/en_us/article/mgbyvb/watch-these-cute-robots-struggle-to-become-self-aware).

337. Bruce Duncan—Talks with the World's Most Sentient Robot, Bina 48," YouTube video, 13:55–17:32, uploaded by ideacity on August 31, 2015, last accessed January 17, 2018, (https://www.youtube.com/watch?v=mwOFWABbfW8).

338. Rachel O'Donoghue, "Google Supercomputer Creates its own 'AI Child,'" December 5, 2017, *Daily Star*, last accessed January 19, 2018, (https://www.dailystar.co.uk/news/latest-news/664713/google-artificial-intelligence-ai-child-NASNet-AutoML).

339. Christina Maza, "Saudi Arabia Gives Citizenship to a Non-Muslim, English-Speaking Robot," October 26, 2017, *Newsweek*, last accessed January 17, 2018, (http://www.newsweek.com/saudi-arabia-robot-sophia-muslim-694152).

340. Tom and Nita Horn, *Forbidden Gates*, 126–127.

341. Hugo de Garis, as summarized in his paper: "The Artilect War: Cosmists vs. Terrans: A Bitter Controversy Concerning Whether Humanity Should Build Godlike Massively Intelligent Machines," *Artificial General Intelligence Conference*, last accessed January 16, 2018, (https://agi-conf. org/2008/artilectwar.pdf).

342. Ibid., emphasis added.

343. David Nield, "Scientists Have Just Made Sperm Out of Human Skin Cells," April 29, 2016, *Science Alert*, last accessed January 20, 2018, (https://www.sciencealert.com/ scientists-create-human-sperm-from-skin-cells).

344. Ibid.

345. https://www.gizmodo.com.au/2018/01/ china-has-already-gene-edited-86-people-with-crispr/

346. David Levy, "An AI Expert Explains How Robot-Human Offspring Would Work," December 22, 2017, *Quartz*, last accessed January 20, 2018, (https://qz.com/1164020/ai-expert-david-levy-author-of-love-and-sex-with-robots-explains-how-robot-human-offspring-would-work/).

347. (https://www.theguardian.com/science/2010/may/20/ craig-venter-synthetic-life-form).

348. Ibid.

349. (https://www.chemistryworld.com/news/wanted-synthetic-chemists-humans-need-not-apply/3008401.article)

350. "L. James Lee and Wexner Medical Center Develop Breakthrough 'One-Touch Healing' Nanochip," August 7, 2017, *OSU: Department of Chemical and Biomolecular Engineering*, last accessed January 20, 2018, (https://cbe.osu.edu/news/2017/08/l.-james-lee-and-wexner-medical-center-develop-breakthrough-one-touch-healing-nanochip).

351. (https://qz.com/1164020/ai-expert-david-levy-author-of-love-and-sex-with-robots-explains-how-robot-human-offspring-would-work/).

352. *Institute for Responsible Technology,* (http://www.responsibletechnology.org/ GMFree/Home/index.cfm).

353. Waclaw Szybalski, *In Vivo and in Vitro Initiation of Transcription*, 405. In A. Kohn and A. Shatkay (eds.), *Control of Gene Expression,* 23–24, and Discussion 404–405 (Szybalski's concept of Synthetic Biology), 411–412, 415–417 (New York: Plenum, 1974).

354. "First Self-Replicating Synthetic Bacterial Cell," *J. Craig*

Venter Institute, (http://www.jcvi.org/cms/research/projects/
first-self-replicating-synthetic-bacterial-cell).

355. Peter E. Nielsen, "Triple Helix: Designing a New Molecule of Life,"
Scientific American (12/08) (http://www.scientificamerican.com/article.
cfm?id=triple-helix-designing-a-new-molecule&ec=su_triplehelix).

356. Charles W. Colson, *Human Dignity in the Biotech Century* (Downers
Grove, IL: InterVarsity, 2004) 8.

357. C. Christopher Hook, *Human Dignity in the Biotech Century* (Downers
Grove, IL: InterVarsity, 2004) 80–81.

358. Garreau, Radical Evolution, 116.

359. Francis Fukuyama, *Our Posthuman Future: Consequences of the
Biotechnology Revolution* (New York: Picador, 2002) 6.

360. Garreau, 106.

361. Garreau, Radical Evolution, 113–114.

362. "Carried Away with Convergence," New Atlantis (Summer
2003) 102–105, http://www.thenewatlantis.com/publications/
carried-away-with-convergence.

363. Summer 2003 issue of The New Atlantis, (http://www.thenewatlantis.
com/publications/carried-away-with-convergence).

364. Bill Joy, "Why the Future Doesn't Need Us," *Wired* (April 2000) (http://
www.wired.com/wired/archive/8.04/joy.html), emphasis added.

365. Mark Walker, "Ship of Fools: Why Transhumanism Is the Best Bet to
Prevent the Extinction of Civilization," Metanexus Institute (2/5/09)
(http://www.metanexus.net/magazine/tabid/68/id/10682/Default.aspx).

366. Ashlee Vance, "The Lifeboat Foundation: Battling
Asteroids, Nanobots and A.I." *New York Times*
(7/20/10) (http://bits.blogs.nytimes.com/2010/07/20/
the-lifeboat-foundation-battling-asteroids-nanobots-and-a-i/).

367. Ian Sample, "Global Disaster: Is Humanity Prepared for the Worst?"
Observer (7/25/10) (http://www.guardian.co.uk/science/2010/jul/25/
disaster-risk-assessment-science).

368. Ibid.

369. Rory Cellan-Jones, "First Human 'Infected with Computer Virus,'" *BBC
News* (5/27/10) http://www.bbc.co.uk/news/10158517.

370. Hook, 92.

371. Ibid., 93.

372. Mihail Roco and William Sims Bainbridge, ed. *Converging Technologies*

for Improving Human Performance (New York: Kluwer Academic, 2003) emphasis in original.

373. Garreau, 7–8.

374. Robert Jeffress, *The Divine Defense: Six Simple Strategies for Winning Your Greatest Battles* (Colorado Springs, CO: WaterBrook, 2006) 78.

375. John Horgon, "We're Cracking the Neural Code, the Brain's Secret Language," *Adbusters* (1/25/06) (https://www.adbusters.org/the_ magazine/63/Were_Cracking_the_Neural_Code_the_Brains_Secret_ Language.html).

376. "Brain-Computer Interface Allows Person-to-Person Communication Through Power Of Thought," *ScienceDaily* (10/6/09) (http://www. sciencedaily.com/releases/2009/10/091006102637.htm).

377. Charles Ostman, *The Internet as an Organism: Emergent Human / Internet Symbiosis* (Vienna, Austria: Thirteenth European Meeting on Cybernetics and Systems Research at the University of Vienna, April 9–12, 1996) emphasis added.

378. Sounding the Codes (2007) http://www.thecenteroflight.net.

379. Robert Maguire, Charles J. Doe, *Commentary on John Bunyan's The Holy War* (2009) 11.

380. Ibid., 7.

381. Frank Jordas, "Study on Cell Phone Link to Cancer Inconclusive," *Associated Press* (2010) (http://abcnews.go.com/print?id=10668283).

382. "How Technology May Soon 'Read' Your Mind," *CBS 60 Minutes* (June 2009) (http://www.cbsnews.com/stories/2008/12/31/60minutes/ main4694713.shtm).

383. "Remote Control Brain Sensor," BBC (November 2002) (http://news.bbc. co.uk/2/hi/health/2361987.stm).

384. R. Douglas Fields, "Mind Control by Cell Phone," *Scientific American* (May 2008) (http://www.scientificamerican.com/article. cfm?id=mind-control-by-cell).

385. Liane Young, Joan Albert Camprodon, et al, "Disruption of the Right Temporo-Parietal Junction with Transcranial Magnetic Stimulation Reduces the Role of Beliefs in Moral Judgments," Proceedings of the National Academy of Sciences (March 2010) (http://www.eurekalert.org/ pub_releases/2010-03/miot-mjc032510.php).

386. Mark Baard, "EM Field, Behind Right Ear, Suspends Morality," *Sci-Tech Heretic* (March 2009) (http://heretic.blastmagazine.com/2010/03/ em-field-behind-right-ear-suspends-morality).

387. Stephen King, *Cell* (New York: Simon and Shuster, 2006).
388. Jeremy Hsu, "Video Gamers Can Control Dreams, Study Suggests," *LiveScience* (5/25/10) (http://www.livescience.com/culture/video-games-control-dreams-100525.html).
389. Aaron Saenz, "Is the Movie 'Inception' Getting Closer to Reality?" (7/15/10) (http://singularityhub.com/2010/07/15/is-the-movie-inception-getting-closer-to-reality-video/).
390. Naomi Darom, "Will Scientists Soon Be Able to Read Our Minds?" Haaretz (http://www.haaretz.com/magazine/week-s-end/will-scientists-soon-be-able-to-read-our-minds-1.291310).
391. Sharon Weinberger, "The 'Voice of God' Weapon Returns," *Wired* (12/21/07) (http://www.wired.com/dangerroom/2007/12/the-voice-of-go/).